SATELLITE BROADCASTING

The Royal Institute of International Affairs is an unofficial body which promotes the scientific study of international questions and does not express opinions of its own. The opinions expressed in this publication are those of the authors.

SATELLITE BROADCASTING

ABRAM CHAYES
JAMES FAWCETT
MASAMI ITO
ALEXANDRE-CHARLES KISS
and others

Published for
THE INTERNATIONAL BROADCAST INSTITUTE
and
THE ROYAL INSTITUTE OF
INTERNATIONAL AFFAIRS
by
OXFORD UNIVERSITY PRESS, LONDON
1973

Oxford University Press, Ely House, London W. 1
GLASGOW NEW YORK TORONTO MELBOURNE WELLINGTON
CAPE TOWN IBADAN NAIROBI DAR ES SALAAM LUSAKA ADDIS ABABA
DELHI BOMBAY CALCUTTA MADRAS KARACHI LAHORE DACCA
KUALA LUMPUR SINGAPORE HONG KONG TOKYO

ISBN 0 19 214994 6

© Royal Institute of International Affairs 1973
    International Broadcast Institute, Ltd., 1973

All rights reserved. No part of this publication may be reproduced, stored in a retrieval system, or transmitted, in any form or by any means, electronic, mechanical, photocopying, recording or otherwise, without the prior permission of Oxford University Press

*Printed in Great Britain by
The Bowering Press Plymouth*

CONTENTS

Preface	vii
Abbreviations not explained in the text	viii
I INTRODUCTION	1
II THE QUESTIONNAIRE	21
III REPLY OF THE FRENCH GROUP	24
IV REPLY OF THE JAPANESE GROUP	67
V REPLY OF THE UNITED KINGDOM GROUP	87
VI REPLY OF THE UNITED STATES GROUP	102
Appendix of documents	134
List of Terms	154
Select Bibliography	155

PREFACE

In 1970 the International Broadcast Institute (IBI) sponsored a four-country legal study of satellite broadcasting. The conduct of the study owed much to the energy and inspiration of Arthur Morse, Executive Director of the Institute, and his untimely death has greatly saddened its completion.

The study has been prepared in France, Japan, the United Kingdom, and the United States. Four persons were chosen by the IBI for their general expertise and knowledge to act, in effect, as rapporteurs: Professor Alexandre-Charles Kiss, of the Institute of Political Studies, University of Strasbourg; Professor Masami Ito, Dean of the University of Tokyo Law School; James Fawcett, Director of Studies, Royal Institute of International Affairs, London; and Professor Abram Chayes, of Harvard University Law School.

In formulating the questionnaire and in preparing the replies, the rapporteurs consulted a number of persons, on a group basis or directly, to ensure that each was reflecting opinions held in his own legal community. The rapporteurs also met from time to time and had the benefit of the advice and experience of Professor Svante Bergström, of the University of Upsala Law School. But the conclusions remain the responsibility of the rapporteurs, and the study does not represent governmental positions.

In order to provide cross-national comparisons, it was decided that the study should be based on a questionnaire formulated by the rapporteurs and comprising the replies to the questions prepared in each country. Despite slight repetitions in the national contributions, it was decided to maintain the original order of the material. Professor Chayes contributed a general introduction, covering developments up until the autumn of 1971. New developments have since taken place and the introduction has consequently been amended to include later information; this work was done by Edward Ploman, presently Executive Director of the Institute. There is also included a select bibliography, and in an appendix the text of a number of pertinent documents. James Fawcett was responsible for assembling the various parts of the study.

September 1972

ABBREVIATIONS NOT EXPLAINED IN THE TEXT

BBC   British Broadcasting Corporation
CCIR   International Radio Consultative Committee
EBU   European Broadcasting Union
ESRO   European Space Research Organization
FCC   Federal Communications Commission (US)
IBI   International Broadcast Institute
IFRB   International Frequency Registration Board
Intelsat   International Telecommunications Satellite Consortium
ITU   International Telecommunication Union
kc/s   kilocycles
MHz   megahertz (hertz=a unit of frequency equal to 1 cycle per second; megahertz=a million cycles per second)
OIRT   Organisation Internationale de Radiodiffusion et Télévision
ORTF   Office de Radiodiffusion-Télévision Française
Unesco   United Nations Educational, Scientific and Cultural
WARC   World Administrative Radio Conference for Space Telecommunications, Geneva 1971
WIPO   World Intellectual Property Organization

I INTRODUCTION Abram Chayes

1. EXISTING SATELLITE COMMUNICATIONS SYSTEMS

In a satellite communications system, satellites in orbit provide links between terrestrial stations sending and receiving radio signals. An earth station transmits the signal to the satellite, which receives it, amplifies it and relays it to a receiving earth station. At the frequencies involved, radio waves are propagated in straight lines, so that in order to perform its linking and relay functions, the satellite must be 'visible'—that is, above the horizon—at both the sending and receiving earth stations during the transmission of the message.

There are at present two different types of systems by which satellites are so positioned: 'synchronous' and 'random orbit'. A satellite placed in orbit above the equator at an altitude of 22,300 miles (35,000 km) will orbit the Earth once every 24 hours. Since its speed is equal to that of the Earth's rotation, it will appear to hang motionless over a single spot on the Earth's surface. Such a satellite is called a synchronous satellite, and the orbit at 22,300 miles above the equator is known as 'the geostationary orbit'. A synchronous satellite is continuously visible over about one-third of the Earth (excluding extreme northern and southern latitudes). Thus a system of three such satellites, properly positioned and linked, can provide coverage of the entire surface of the Earth, except for the arctic and antarctic regions.

A satellite in any orbit other than a synchronous orbit will be simultaneously visible to any given pair of earth stations for only a portion of each day. In order to provide continuous communication between such stations, more than one satellite would be required, orbiting in such a way that when the first satellite disappeared over the horizon from one station, another had appeared and was visible to both sending and receiving earth stations. The number of such satellites required to provide continuous communication depends on the angle and altitude of the orbit. The number could be minimized if the spacing between the satellites were precisely controlled (controlled-orbit system), but a somewhat larger number with random spacings can achieve the same result (random-orbit system).

Since the synchronous satellite remains stationary with respect to any earth station, it is relatively simple to keep the antennas at the sending and receiving stations properly pointed at the satellite. Only small corrections for orbital errors are required. In a random-orbit system, the earth-station antenna must track the satellite across the sky. Moreover, if continuous communication is to be maintained, a second antenna must be in readiness at each earth station to pick up the following satellite as the first one disappears over the horizon.

At present, there are operating systems of both main types. Intelsat operates a synchronous system providing global coverage, with satellites positioned above the Atlantic, Pacific, and Indian Oceans. The Soviet Orbita system, used for space network domestic communications (including television distribution) within the USSR, is a random-orbit system, using eight satellites and providing continuous 24-hour communications. The satellites are spaced so that two of them are always over the Northern Hemisphere; and the orbits are such that during the time when it is in operation, a satellite is at the apogee of its orbit. Its apparent motion with respect to the Earth's surface is slowest at this time and the tracking problem is minimized.

Satellites in existing systems of both types generate relatively low effective power, in part because of payload limitations on launch vehicles and in part because satellites, at least of the early generations, diffuse their power around a sphere instead of focusing it in a single direction. Most of the power generated by such a satellite is radiated into space instead of being concentrated at the point on the earth where the message is to be received. In order to pick up the relatively faint signal from these satellites, sensitive and correspondingly costly earth-station antennas are needed. The Intelsat standard earth station costs from $3–5 million, while the Orbita stations are reported to be less costly. Tracking facilities for random-orbit systems are also expensive. Costs can no doubt be reduced considerably even with present satellite technology, but multiple reductions will require satellites with substantially increased effective power.

In part because of the high cost of earth stations, all satellite communications systems to date, whether of telephone and telegram message or television programs, have been point-to-point systems. Messages are gathered and transported to the sending earth station by conventional means of communication, primarily coaxial cable and microwave relay; and at the receiving end the signal is taken from the receiving station, again by land or microwave links, to existing terrestrial telecommunications or broadcasting transmitters, which, in turn, transmit to the user or to home receiving sets.

II. BROADCAST SATELLITES

Broadcasting of radio and, especially, television programs via satellite has been an exciting prospect from the earliest discussions of satellite communications. For radio broadcasting, the use of satellites could improve the quality of the signal in long-distance transmissions, and reduce the transmission to a single stage; and, because existing conventional systems broadcast over very long ranges, the introduction of a satellite repeater would not create significant new problems in terms of technical regulation or control of program content. The international community has been living for some decades with the problems of interference, propaganda, and protection of program content created by the ability of national broadcasting entities to transmit their programs at long range across state lines.

For television, however, the advent of the broadcast satellite has an entirely different significance. The propagation characteristics of television signals confine their reach to a short radius, perhaps fifty miles around the transmitters, in effect a line-of-sight limit. To send programs beyond that range requires that these essentially local broadcasting stations be linked by some terrestrial facility—coaxial cable, microwave relay, or by the dispatch of film or videotapes to transmitters by mail. Because of these technological limits, television broadcasting to date has been largely a domestic affair unless special arrangements are made, as in the case of Eurovision and Intervision. International problems created by harmful interference or offensive program content have not been acute, are confined to areas along the boundary between two states, and are thus primarily bilateral in character. These insulating conditions dissolve with the introduction of satellite repeaters in synchronous orbit. With such a system, it would be possible for the first time for a single sending station to broadcast its television programs around the globe, without the interposition of any externally-controlled linkages between it and the terrestrial receivers. The synchronous satellite opens up the same capability for television broadcasting that has existed until now for powerful short-wave radio transmitters, with the added factors of the more decisive impact of the television medium and the much higher quality of the satellite signal.

The ultimate development of broadcast satellites would be to use them for direct broadcasting to ordinary home receivers, and much of the public discussion has focused on this possibility. But here the distinction between broadcast satellites and other communications satellites is not hard and fast. In the point-to-point systems described above, the satellite relays a signal to a relatively small number of complex and expensive receiving stations. A broadcast system would involve larger numbers of simpler and relatively inexpensive receivers. In fact, some systems, primarily in the developing countries, will use the same satellite for both point-to-point message communications and for radio and television broadcasting.

The Working Group on Direct Broadcast Satellites of the UN Outer Space Committee has distinguished three different types of broadcast-satellite systems:

(1) 'direct broadcasting television signals into existing, *unaugmented* home receivers . . .'
(2) 'direct broadcast of television into *augmented* home receivers'.
(3) 'direct broadcast into community or collective receivers'.

These categories correspond roughly to the definitions of three grades of broadcast satellite services proposed by the CCIR of the ITU:

Primary (Principal) Grade of Service
A grade of service such that a power-flux density of sufficient magnitude to enable the general public to receive transmissions directly from the satellites by means of individual installations and with a quality comparable to that provided by a terrestrial transmitter to its primary service area.

Secondary (Rural) Grade of Service
A grade of service with a lower power-flux density than that required for a primary grade of service, the signals of which are intended for direct public reception from satellites by means of individual installations and with an acceptable quality in sparsely populated areas which are not served, or are inadequately served, by other means and where satellite reception conditions are favourable.

Community Grade of Service
A grade of broadcasting service from satellites with a limited power-flux density, the signals of which are intended for group viewing or listening or for reception by a master receiver installation.

The ITU as the UN Agency responsible for telecommunications matters, adopted at the World Administrative Radio Conference for Space Telecommunications (WARC–ST) in 1971, the following definitions for satellite broadcasting.

Broadcasting—Satellite Service
A radiocommunication service in which signals transmitted or retransmitted by space stations are intended for direct reception by the general public.

This definition has been complemented with a footnote stating:

'In the broadcasting-satellite service, the term "direct reception" shall encompass both individual reception and community reception.'

Individual Reception (in the broadcasting-satellite service)
The reception of emissions from a space station in the broadcasting-satellite service by simple domestic installations and in particular those possessing small antennas.

Community Reception (in the broadcasting-satellite service)
The reception of emissions from a space station in the broadcasting-satellite service by receiving equipment, which in some cases may be complex and have antennas larger than those used for individual reception, and intended for use:

—by a group of the general public at one location; or
—through a distribution system covering a limited area.

It will be seen that these definitions refer to points on a continuum rather than to physically or technically distinct concepts. Systems already in the advanced planning stages display a mixture of the types defined and such 'hybrid systems' can be expected to be the rule rather than occasional exceptions.

A crucial factor is who controls, in the first instance, the choice whether or not a particular satellite program is to be received. If the satellite can broadcast directly to available home receivers, whether or not 'augmented', that choice is in the hands of the owner of the home television set. But if a link is interposed between the satellite and the ultimate viewer, the control over what programs reach the viewer will be in the entity managing that link. In most countries that entity will be a government agency or licensee. It is irrelevant, for these purposes, whether the link is a community receiver in a school, community center, or market where the population to be served gathers to watch; or whether it is a community antenna from which the signal is distributed to home sets by cable or conventional over-the-air broadcasting. It is likewise irrelevant that the community receiver or antenna may be hardly more complicated or expensive than an 'augmented' home receiver. The critical difference from this point of view is not technical or economic, but organizational. It depends on whether the system is set up so as to interpose some kind of controllable link between the broadcast satellite and the ultimate viewer.

III. PRESENT PROSPECTS FOR BROADCAST-SATELLITE SYSTEMS

A. *Technical and Economic*

Broadcasting of *sound radio* directly from satellites to home receivers is probably feasible with current technology, but such systems have so far attracted little interest perhaps because they do not add enough to terrestrial long-distance radio-transmission capability. Broadcast satellite systems for *television* are still some years in the future. Using the taxonomy discussed above, the UN Working Group, in February 1969, gave the following estimates:

(i) Direct broadcast into community receivers could be close at hand. Technology currently under development might allow this in the mid-1970s. Such a system is considered to be less expensive to launch than one intended to be for reception directly in people's homes. . . .

(ii) Direct broadcast of television into *augmented* home receivers could become feasible technically as soon as 1975. However, the cost factors for both the earth and space segments of such a system are inhibiting factors. Therefore, it is most unlikely that this type of system will be ready for deployment on an operational basis until many years after the projected date of feasibility.

(iii) ... direct broadcasting television signals into existing, *unaugmented* home receivers on an operational basis is not foreseen for the period 1970–1985. This reflects the lack of technological means to transmit signals of sufficient strength from satellites.

These estimates have been borne out on the whole by developments since that time. The system, being developed by India in cooperation with NASA, which will use community receivers in part, is scheduled to go into operation in 1975 on an experimental basis using a NASA experimental satellite. This phase would be followed with an ongoing program using an Indian synchronous satellite, Insat I. Community receivers for the system will cost about $100 over and above the cost of a conventional television set. The figure is below the $150 augmentation cost for community receivers estimated by the UN Working Group in 1969, and suggests that augmentation for home receivers would be available at prices within the reach of viewers in the developed countries before many more years.

Despite these cost reductions, it has become increasingly doubtful whether satellite broadcasting to home receivers, augmented or not, will be widely realized as a practical matter, at least with the present basic technological concepts. The developed countries have extensive land lines and terrestrial broadcast facilities. It seems likely that these will continue to be used to distribute satellite broadcasts from common central antennas to home receivers. In the developing countries, with tight constraints on available resources for many years to come, the cost of providing home receivers for the population at large will remain prohibitive. At its third session, in May 1970, the Working Group noted 'the absence of any evidence that programmes were being pursued to develop the very costly and complex technology of broadcasting from satellites direct to either augmented or unaugmented home receivers ...'. It therefore concluded that there was no need to revise its earlier time estimates. This conclusion seems sound in view of the lead times involved.

Although the prospect seems unlikely, it is not impossible that a state or group of states would try to put up a satellite powerful enough to reach home or community receivers in a foreign area. At present, only two states, the United States and the USSR, have launch capability and have, or can readily obtain, the communications technology necessary for such an effort. Japan and Canada will soon join them. The nations of Western Europe, if they united in the enterprise, and, some day, China, must be considered as potential candidates, but the list cannot be extended much further.

Introduction

Even though such a system is technically conceivable, the technical and administrative rules adopted by the WARC–ST provide not only for notification and co-ordinating procedures prior to putting a satellite broadcasting service into operation; with regard to signals transmitted into adjacent countries unavoidable spill-over is acceptable but in all other cases the prior agreement of receiving countries is required. Moreover, practical considerations make it most unlikely that such systems will ever be realized operationally. The cost could be considerable, even for the United States and USSR. For the other potential operators, such a system must be very far down the list of national priorities. In Europe there would be, in addition, the formidable difficulty of summoning the necessary degree of technical, economic and political co-operation.

Despite the political, administrative and practical considerations which make such systems unlikely, fears have been expressed with regard to two purposes, which have been suggested for the adoption of such a system. The first is propaganda. For the United States and the USSR, the experience of global propaganda efforts by radio and other media after World War II has been discouraging. The cost is high and the results meager. Moreover, there is increasing sensitivity to the political costs of intrusive propaganda programs in target countries. These two factors make broadcast-satellite systems for propaganda purposes an extremely unlikely investment, at least in the absence of conditions of much more intense conflict than presently prevail.

A possibility remains that private commercial interests might be willing to finance such a system as a vehicle for advertising penetration of new markets. At present, the ratio of costs to benefits of such a project would seem excessive in comparison with available alternatives. It is not out of the question that developments over the next decade or so will change this calculus.

It should be noted that legal and administrative rules adopted on the international level make it impossible for private firms to operate such systems without authorization from their national Government.

The absence of any existing development programs looking toward systems of this type, noted by the UN Working Group, make it seem almost impossible that such a system could be operationally realized before some time in the decade of the 1980s—if political and economic considerations would make such a system attractive, which seems unlikely.

B. *Organizational*

For a long time the question of organizational arrangements for satellite communications has been a principal focus of discussion. A number of observers, including a Task Force organized by the

Twentieth Century Fund, favored a comprehensive and integrated international system providing its members with a full range of satellite communications services, including direct broadcasting. The system would have been internationally owned and operated under the aegis of Intelsat, and would ultimately have linked, whether by merger or otherwise, with Intersputnik, the international system proposed by the USSR. Whether or not this was a wise conception, the Plenipotentiary Conference on Definitive Arrangements for Intelsat has not embraced it. It seems very unlikely, therefore, that there will be a comprehensive and integrated system of the type described, at least for a considerable period in the future.

At the opposite pole, one might imagine a set of national or regional satellite broadcast systems, serving primarily domestic needs and audiences, being linked with each other or the global system only on special occasions. Organizationally, satellite broadcasting, in this model, would look much like terrestrial broadcasting today. Indeed, satellite systems might simply be a part of national broadcasting organizations.

It is not impossible that developments will take this course. Almost certainly, large countries like Brazil, Canada, China, the Congo, India, Indonesia, the USSR, and the United States will ultimately have established one or more domestic communications satellite systems. The USSR already has a point-to-point domestic system; the United States and Canada, India and Brazil are all planning satellite communications systems, broadcasting to community or collective receiving installations. For smaller countries, however, it still seems likely that economies of scale, both for hardware and programming, will dictate cooperative efforts to form regional systems among geographically contiguous and/or culturally compatible states. The UN Working Group, in its latest session (May 1970), seems also to envisage this kind of organizational setting. Although in its first two meetings, much of the discussion focused explicitly or implicitly on problems of global broadcast systems, the report of the third session was 'firmly of the opinion that the greatest prospects for practical cooperation are at the regional level.' The group 'foresaw the establishment and operation of regional satellite broadcast systems and program planning and production for such systems on the basis of regional cooperation'.

Although one cannot be certain, it appears that the number of separate systems is not likely to be very great, at least for the immediate future. The Soviet system is in place. India and Canada have proceeded beyond the initial stages of planning, and it can be expected that their systems will be in operation within the decade of the 1970s. The Executive Branch of the US government has pressed for policies hospitable to the establishment of domestic systems, but the appliances filed with the FCC have not yet been fully endorsed. The other possible candidates for national systems—no more than four or five states at most (such as Brazil, Indonesia)—have hardly begun work.

On the regional side, the likeliest prospect is the Spanish-speaking countries of Latin America. Here, planning is in the preliminary stages. A joint request by Argentina, Bolivia, Chile, Colombia, Ecuador, Paraguay, Peru, Uruguay and Venezuela, for a feasibility study of a regional system for education, culture, and development information which would use communication media including broadcast satellites has been accepted by the UN Development Programme. There has been much discussion of the Symphonie project by France and Germany, which could be used for linking France with French-speaking areas in Africa, the Caribbean, and North America.

The West European countries have made detailed examinations of a regional satellite system for both telecommunications and television transmissions. According to the decisions made by ESRO late in 1971, an operational European system is foreseen for about 1980. Regional systems for West Africa, possibly one for East Africa, might make sense at one point, as might one for the Arab region. The Soviet system will probably evolve into or append an international system; at present the Eastern European countries, Cuba, and Mongolia are members of the projected system (Intersputnik). A Japanese-sponsored system may link Southeast Asia. That just about exhausts the possibilities.

The prospect is, thus, for a maximum of some twenty separate systems, global, regional, or national of various kinds, in the next phase of the development of communications satellites. It would be somewhat surprising if as many as half that number were to come into operation before the end of the decade, and not all these will be broadcast types.

Technical and legal factors would operate to make each national or regional broadcast satellite system a discrete, self-contained entity. The community receivers in each system would be designed and set on frequencies so as to permit reception only from a satellite in their own systems. The satellites themselves would also employ frequencies and design characteristics that would prevent interference with adjacent systems. Failure so to construct the system would violate existing ITU rules against harmful interference. *A fortiori*, the deliberate operation of an out-of-system satellite with frequencies and other characteristics designed to reach receivers within the system would violate the ITU obligations of the sending states. Satellite antennas now being developed can focus the transmission from the satellite into a fairly narrow beam falling on a given segment of the

Earth's surface. Designs of this kind will increase the effective power radiated by the satellites and at the same time make it possible to limit spillover of the transmission into territories outside the system.

The internal regime for national or regional systems would be established by the states participating in the system. For national systems, it would be defined by the domestic legislation of the state involved. The UN Working Group has consistently noted that 'in the case of direct broadcasting into community receivers for purely domestic coverage, a Government ... would be able to adopt such regulations as it considered appropriate, and that, in this situation this would lead to lessening of the problems which needed international regulations'. For regional systems, agreement among the participating states would be required to establish technical specifications, financial contributions, operating conditions, time-sharing arrangements, and any necessary regulations for program content. To quote the Working Group once more, 'For regional ... coverage into community receivers, there would appear to be a requirement for regional ... cooperation and coordination in such matters as uses of satellites, common technical standards, languages, time sharing and programme content.' In the report of its third session, the Group suggested that 'regional arrangements could be based on the principle that countries receiving the signal transmitted by the satellite would participate in the planning and preparation of programmes or else in the management and operation of a regional based system'.

This picture of a set of more or less separate national and regional systems seems to negate one of the main advantages claimed for satellite broadcasting, the potential contribution to world understanding involved in the possibility of reaching a world-wide television audience. The notion of a universal audience watching the same programs at the same time is exaggerated, however. A third of the world, at least, is asleep at any given hour. Barriers of language and culture further fragment the potential audience. Only for a few special and universally symbolic events does it seem likely that a global broadcast network would be in demand: the moon landing, coronations and assassinations, the Olympics or the football championships, important messages of heads of state in times of crisis, certain proceedings of the UN, perhaps a few others. Again, the UN Working Group has reached similar conclusions: 'Owing to difficulties raised at the outset by differences in time-zones, varying programme interest and viewing patterns and economic considerations, global broadcasting satellite systems could not be foreseen at present. The wide or global distribution of programmes was expected to remain a rare occurrence, limited to coverage of very special events.'

This limited demand can and will very likely be met by linkage between separate broadcast satellite systems. Today interconnection, even between incompatible systems, can be achieved by a terrestrial link between two earth stations, one in each system. Moreover, two systems could be designed so that a single earth station, properly positioned, could operate in both. This is apparently already possible, subject to the installation of special antennas for working with both systems, as between the Intelsat and Molniya systems. The French earth station of standard Intelsat design has already picked up Molniya transmissions on an experimental basis. In the relatively near future it will be possible to effect a linkage between systems by transmission directly from a satellite in one system to a satellite in the other, without the intervention of a terrestrial sending or receiving station. National and regional systems can and most likely will be designed so that, like existing network arrangements, the link-up will be at the option of the receiving rather than the originating system. This kind of interconnection capability would pose additional but not insurmountable problems of technical coordination. The gateway between systems would provide a point at which regulatory authority could be exercised.

C. *Impact on Existing Communications Systems*

1. *Developed countries*

In the developed countries, with highly evolved terrestrial systems of television broadcasting, satellites are likely to be fitted into the existing network in an auxiliary role, at least for the next decade or so, and will not dominate television broadcasting or fundamentally change its character. In the field of commercial broadcasting in the United States, for example, the principal interest in satellites is for cheap networking facilities to replace coaxial or microwave terrestrial links now in use, and the impact of cable TV is potentially more significant for the structure and organization of the US broadcasting industry, and perhaps for Europe and Japan, as well.

However, the role of satellites is not yet clear, and in their recent applications to the FCC RCA Global Communications offers a 40–50 per cent saving in network TV transmission costs, and Hughes Aircraft seeks a satellite to provide linkage among cable systems. Satellites will be able to reach viewers in remote and sparsely settled regions of the developed countries, where it would be uneconomic to link a local broadcast station to the national network by terretrial means. This function has some significance for the United States and is very important for Canada, Japan, and the USSR. Beyond this, satellites may permit broadcasters in developed countries to do things they are already doing with added convenience or economy, and immediacy, in the case of distant events.

In Western Europe present frequency assignments preclude more than about four national television programs in each country; so far, no country has developed more than three programs. However, viewers in border areas seem willing to pay for sets that will receive programs broadcast in neighbouring states. This suggests that the demand for programs is not saturated. In these circumstances, a regional satellite system operating on narrow beams directed towards single countries or groups of countries might provide facilities for further television programs.

Developing countries
It is in these countries, without extensive television broadcast systems already in place, that the revolutionary potential for direct broadcasting most clearly appears. With broadcast satellites, a full-blown national television capability can be established in these areas at relatively modest cost. Moreover, in countries of wide geographical extent, the same satellites may provide links for inter-city point-to-point communications.

In the developing countries the satellite-broadcast system is likely to be under direct governmental control. Broadcasting is conceived as a state enterprise throughout much of the developing world. In addition, the costs of the space segment—even a share of the costs of a regional system—are still high enough to require public financing. Finally, as already noted, the economies of the less-developed countries will not support widespread private ownership of home receivers. Thus receivers are likely to be supplied by the state to designated communities or institutions, and for specified purposes. For both economic and political reasons, these receivers are likely to be simple installations, capable of receiving only the channels, perhaps a single channel, on which the state broadcasting entity operates.

In such circumstances, access to the system and control over programming will be firmly in the hands of the state and subject to its schedule of priorities. In all less-developed countries, education is close to the top of this priority list. Education in a prescribed curriculum, besides its intrinsic value, will contribute a sense of national identity and national unity.

The educational objectives of developing countries are unattainable by conventional methods within an acceptable time period. Satellite broadcasting to schools through the country holds large promise of filling this gap, particularly in the important areas of literacy, training in basic agricultural and mechanical skills, population control, and the like. The broadcast satellite systems now closest to realization are primarily educational systems.

On the other hand, satellite broadcasting will levy claims on some of the scarcest resources of developing countries. In particular, the engineering and technical skills needed for design and operation of such systems are likely to be in desperately short supply, and urgently needed for all the tasks of development. Some components and perhaps simpler community receiving installations, may be suitable for manufacture or assembly in countries in the early stages of industrialization. The satellites themselves, however, as well as launch services and, in all likelihood, the complicated transmitting stations will have to be procured abroad at the expense of precious foreign exchange. Requirements for power supply, maintenance work, and operating personnel in remote and inaccessible parts of the country may distort development priorities. Moreover, many countries have only limited, if any, indigenous program production capability, so that software, too, may have to be purchased abroad. In addition to the economic costs, this will carry with it all the sensitive problems of translation and adaptation of educational materials from one language and cultural setting to another. A special problem will be to avoid programming that depicts, even unconsciously, affluent Western life-styles, thus stimulating demand for import of consumer goods in general. Teachers, too, must get special training to participate effectively in education by television (although it may well be that the broadcast satellite system itself can serve the urgent teacher-training needs to be found in most developing countries). When all is said, and although there have been promising experiments, educational television in general has not yet lived up to its advance billing. Predictions of its effectiveness in developing countries should probably be taken with some reserve, at least for the short range. These considerations suggest that satellite broadcasting is by no means the panacea for the educational and cultural needs of developing countries. Costs and benefits of specific proposals must be carefully evaluated in terms of the conditions and the overall development plans of the particular country involved.

The Indian system. Now nearing operation on an experimental basis, the Indian system illustrates most of the technical, economic and political features of separate satellite broadcasting systems noted above. A review of that system in some detail may be helpful to give a more concrete sense of what is to be expected in satellite broadcasting in developing countries over the next decade or so.

Indian planning for satellite communications began as early as 1962–3. The first step in the program was the establishment of an experimental satellite earth station, in collaboration with the UN Development Program and the ITU. The primary purpose of this installation was to begin to develop an experienced cadre of Indian operating engineers and technicians. The project was begun in 1965 and completed in 1967.

In September 1969 arrangements were concluded with NASA to

carry the experiment forward on a larger scale. The Satellite Educational Television Experiment provided for by this agreement will reach 5,000 hitherto isolated Indian villages with instructional programs in agricultural productivity, general education, and population control. The Experiment will use the NASA ATS-F satellite, to be launched in 1974. After a year of tests and other experiments, the satellite will be moved into position over the equator south of India, where it will remain for a year. The satellite will be able to focus its beam within a subtended arc of 3°, providing the effective power necessary for a broadcast signal and confining the transmissions primarily to the Indian subcontinent. The satellite will provide one video and two audio channels to permit simultaneous transmission in two different languages.

Two forms of reception will be employed. In certain areas that will have conventional TV broadcasting facilities—e.g. Delhi, Bombay, and Srinagar—the signal will be picked up from the satellite by an earth station and relayed to the conventional transmitter for rebroadcast to community receivers in villages within that station's broadcast area. Three thousand of the 5,000 villages to be covered will be reached by this method. The other 2,000 villages will have community receivers. These receivers will consist of an antenna made of chicken wire to receive the signal directly from the satellite, and a converter to present this picture on the screen of a conventional television set in a schoolroom or village center. This will be the first example of direct broadcasting from a satellite to community receivers without the intervention of relay stations on the ground. Both the antennas and converters for these receivers will be manufactured in India, creating investment and employment opportunities and conserving scarce foreign exchange. Cost estimates on the antenna run to about $25 each, and for the converter, $75 each, amounting to $100 per village, in addition to the cost of a conventional receiving set. The villages to be covered by this means are clustered in some of the least developed parts of India.

Plans are already in train for putting the system on a permanent basis after the end of the experimental period with the projected launch in 1975 of Insat I, a communications satellite to be owned and operated by India. Coverage will be extended to an additional 100,000–150,000 villages per year, achieving all-India coverage by 1980. In addition to its broadcast capability, the permanent system will provide point-to-point connections for message traffic between major Indian population centers, which are currently without satisfactory terrestrial telecommunications links.

Total system costs are estimated in the range of $200 million or $40 million annually over the five-year period. It is estimated that to achieve that same coverage by conventional means, and the

Introduction

equivalent annual expenditures, it would take until 1999 at a total cost of $1,160 million.

The first prototype of Insat I will be built in the United States, but with the participation of Indian engineers and technicians. It is hoped that within the same 1975–80 period India will be able to design and build her own satellites. Present projections also contemplate the development of an indigenous launch capability in the same period.

Meanwhile intensive work is going forward on programming. Indian authorities will be solely responsible for the content of broadcasts, both during the experimental period and thereafter. Programs will be developed in India, using the large existing motion picture industry as a base. Elaborate plans are already under way for the cooperation of academic specialists, government departments, and technical experts in program development and evaluation.

IV. MANAGEMENT OF SPECTRUM AND ORBIT RESOURCES

A. *Technical Considerations*

Communication by satellite depends upon the use of a portion of the radio-frequency spectrum. Voice, message, and data transmissions occupy only small band-widths in the spectrum: for example, one voice circuit requires theoretically only 4 kc/s of bandwidth, and in practice can be realized with 6–8 kc/s. Television transmission, however, alone requires 6 MHz (6,000 kc/s) of bandwidth. The prospect of a number of separate systems engaged in satellite broadcasting therefore raises the question of whether existing frequency allocations, and others that might be made, in desirable portions of the spectrum will be adequate to accommodate the likely demand.

The availability of frequencies is not the only variable in the equation, however. Both sending and receiving earth stations are designed with their antennas pointed at the satellite in their own system. If that satellite is spaced sufficiently far from any other satellite transmitting on the same frequencies, the system can operate without harmful interference from the second satellite. Their required spacing is sensitive to a number of factors, including principally satellite power and the size and design of earth-station antennas. For synchronous systems using earth stations with antennas 30 feet in diameter, presently available technology dictates an interval between satellites of about 5°. If no other variables were involved, this would permit deployment of 72 separate satellites using the same frequencies around the 360° of the synchronous orbit.

This represents far more than the number of communications satellites, including broadcast satellites, projected for deployment in the coming decade.[1] However, not all portions of the orbit are equally suitable for broadcast satellites. For these purposes, as well as for other types of domestic or regional synchronous communications satellites, the arc over the equator directly north or south of the area to be served is most valuable. For practical purposes these portions fall between the longitudes of the major continental land masses.

For example, the area of greatest interest for satellite broadcasting to the United States lies between 60° and 135°W longitude. Satellites within this sector will be visible throughout the continental United States and in the western 20° will reach Hawaii and Alaska as well. With 5° spacing, this portion of the arc would accommodate 16 satellites, a number still far in excess of presently projected US demand for domestic broadcast and other satellite-communication services. This same portion of the orbit, however, is also optimal for Canada and Mexico, and the eastern segment contains the most desirable locations for Latin America. If 5° spacing were indeed a limit it is not hard to see the possibility of orbital congestion, if not in the next decade, then at some time thereafter if systems continue to proliferate as demand for communications channels increases.

It is not true, however, that the 5° spacing is a fixed constraint on the number of operating satellites that can be placed in the synchronous orbit at any one time. First, as already noted, this amount of spacing is required only for satellites using the same frequencies. For satellites using different frequencies, the spacing can be reduced to the minimum necessary to ensure that the satellites will not collide —about 2° with present control techniques. Of the frequencies now allocated for space communications only a part is presently used for commercial communications, although other portions have been reserved by some states for military and government use in some areas of the world. By using the remainder of the presently allocated frequencies, the capacity of the synchronous orbit could be at least doubled. The possibilities of allocating additional frequencies, thus further expanding total communications capacity, was one of the principal items on the agenda of the 1971 WARC of the ITU.

Even if satellites use the same frequencies, a variety of other design features, available with existing technology, can operate to expand the capacity of the spectrum/orbit resources to accommodate additional satellite communications systems. A few of the more prominent of these may be mentioned briefly:

1. Satellites can now be made to focus their transmissions on a relatively limited area, ultimately, as some believe, as small as a few

[1] Satellites providing other than communications services—e.g. meteorological, navigational, geodetic—will not change the picture substantially, since they will probably not use the same frequencies, nor, in many cases, the same orbits as communications satellites.

Introduction

hundred miles in diameter, and perhaps smaller. Two satellites, even though directly adjacent and using the same frequencies, will not interfere with each other so long as the areas on which the two beams focus do not overlap. Thus, for example, a satellite with its beam focused on the United States would not interfere with an adjacent satellite using the identical frequencies, with its beam focused on South America. In fact, a single satellite could readily be designed to transmit both beams. The ability to focus beams more narrowly is increasing rapidly. The narrower the beam, the smaller the spacing required between satellites in orbit, until the practical limit is reached—the spacing needed to prevent collision between satellites.

2. Today in the Intelsat system all transmissions between Earth and satellite use a width of 500 MHz in two portions of the frequency spectrum, 3,700–4,200 MHz for Earth-satellite transmissions and 5,925–6,425 MHz for satellite–Earth transmissions. This pattern is necessary because a single satellite system cannot transmit and receive on the same frequency at the same time without interference. But if a second satellite system were also broadcasting, it could reverse the frequencies, using the higher frequency band for the up-link and the lower for the down-link. This would effectively double the capacity of the orbital arc to accommodate common-frequency satellites.

3. Additional possibilities for increasing the capacity of the orbit with presently available technology include antenna polarization, the type and degree of modulation used to impress the messages on the radio waves and increasing the proportion of interference permitted within the total noise allowances fixed by ITU Regulations.

B. *Legal and Institutional Setting*

In the light of these variables—and technological improvement will produce more of them—it seems likely that the spectrum/orbit resources are potentially adequate to meet almost any demand at present conceivable. In practice, however, this potential abundance cannot be realized without intricate and complex coordination among separate systems, beginning with the earliest stages of planning and design. What existing legal and institutional machinery is available to perform this necessary coordination function?

Frequencies are now allocated for various international telecommunications services by the ITU, acting through periodic Administrative Radio Conferences. The decisions of the Conferences are not binding on any government until they are adopted through its normal processes for treaty ratification, which may in some case require legislation or other similar action. Once a portion of the spectrum is so allocated—in this case for satellite-communications services—frequencies within the allocated portion are assigned to particular operating systems by the national communications administration concerned. The ITU Convention establishes a procedure for registration of these frequency assignments by an agency of the ITU, the IFRB.

Until the WARC of 1971, registration, in the case of satellite-communications systems, required notice of the frequencies to be used, the proposed orbital position and certain other characteristics, notably the effective power at which the satellite would operate, antenna directionality, and other matters relevant to compliance with the criteria established for use of frequencies already in use by terrestrial services in the area of coverage. If these characteristics were in conformity with ITU Regulations, and if there were no likelihood of interference with stations already registered, the applicant would be entitled to have the frequencies registered in the Master Register with a favorable finding. That meant, in effect, that the system was entitled to priority over any later systems that caused interference with it, even though the registered system was not designed so as to economize spectrum use and did not take account of prospective needs of other users in its planning.

The 1971 WARC made some changes in this framework. Most important, it required that new satellite systems should coordinate not only with existing terrestrial stations, but with existing satellite systems as well. The coordination process has been broadened somewhat to give prospective users of the spectrum, as well as existing users, some standing to object. And existing systems are placed under an obligation to *consult* with a newcomer about adjustments to accommodate him, though there is no substantive requirement that any adjustment be made. Finally, the WARC issued a call for a planning conference for broadcast satellites, and adopted a set of interim regulations that are relatively inhospitable to broadcast systems. Overall, the coordinating process remains one for a series of bilateral adjustments of national policies rather than an integrated spectrum-management function. The regime of first-come-first-served is hardly altered.

On the procedural side, the IFRB remains without the technical capability and the legal authority to make an independent investigation of an applicant's compliance with the regulations. It is not empowered to deny registration, even if it makes an unfavourable finding. And it cannot adjudicate disputes over harmful interference and enforce its decisions.

Issues relating to the legal regime for satellite communications and to the organization and procedures of the IFRB as well as broader issues concerning ITU structure and powers will be before the ITU Plenipotentiary Conference in 1973.

V. OTHER DEVELOPMENTS

Apart from the UN and ITU, action on the international, intergovernmental level has also been taken by Unesco. The Unesco activities have mainly focused on two aspects.

First, studies concerning the use to which communications satellites can be put with regard to the flow of information, education, and culture. In this respect, Unesco has, at the request of the respective governments, sent expert missions to a number of countries for preliminary studies on the possible use of satellites for education and national development. Such missions have visited India, Brazil, Pakistan, Spanish-speaking countries in South America, and a number of Arab States; studies are also going on in Africa.

Second, Unesco has, following a number of expert meetings, drawn up a 'Draft Declaration of Guiding Principles on the Use of Satellite Broadcasting for the Free Flow of Information, the Spread of Education and Greater Cultural Exchange'. This Draft Declaration is to be put before the 1972 session of the Unesco General Conference. The Declaration is intended to cover areas falling within the mandate of Unesco and sets down standards without taking the form of a binding international agreement.

It should be noted that in August 1972, the USSR requested for inclusion in the agenda of the 1972 session of the UN General Assembly 'Preparation of an international convention on principles governing the use by States of artificial earth satellites for direct television broadcasting'. This proposal is for a binding international instrument. The main provisions concern the equal rights of all States to carry television broadcasts via satellites; the applicability of generally recognized principles of international law, including the UN Charter and the 1967 Outer Space Treaty to such broadcasts; prohibition of TV broadcasts via satellites to foreign States save with their express consent. A number of rules concerning the content of satellite broadcasts are included. It is also proposed that a State may counteract illegal satellite broadcasts not only in their own territory but also in outer space.

Finally, mention should also be made of the work carried out by Unesco and WIPO with regard to legal problems raised by satellite transmissions in the field of copyright and 'neighbouring' rights. A main purpose of this work has been to specify whether the protection of television signals in themselves transmitted by communication satellites requires modification of existing conventions or the preparation of a new international instrument.

A committee of government experts has held two meetings in 1971 and 1972. A draft text for a new and independent treaty has been examined but since there is still a lack of broad consensus with regard to a number of provisions in this draft, study of this matter will continue and it is expected that Unesco and WIPO will convene a third meeting of the Committee of experts.

II THE QUESTIONNAIRE

1. What means of promoting and extending freedom of information does direct satellite broadcasting provide, in particular, for developing countries?
2. In what ways, institutional or normative, may
   a. hostile propaganda
   b. programs tending to disturb or erode religious, cultural, or social life
   c. advertising

   be limited or prevented in direct satellite broadcasting consistent with the principle of freedom both to impart and receive information?
3. In what ways does the individual require protection in direct satellite broadcasting from
   a. defamation or false statements in particular by a right of reply and the duty of rectification
   b. invasion of privacy such as undue publicity of personal life?
4. In what ways, institutional or normative, do authors, performers and producers of broadcast programs, and producers of phonograms, need special protection for contributions to direct satellite broadcasts in respect of
   a. economic interests in the contribution
   b. the intellectual and artistic integrity of the contribution?

   Further, do satellite broadcasts themselves need special protection?
5. What problems do the issues raised in question 4 create for developing countries in the reception of direct satellite broadcasts or access to satellites?
6. How is direct satellite broadcasting to be correlated with other systems of satellite communications and in particular
   (i) What problems does direct satellite broadcasting create in respect of the determination of orbital positions of satellites, the allocations of frequencies, measures to avoid interference with other transmissions, and measures to limit or prevent spill-over, to be used in direct satellite broadcasting for the solution of the problems considered in questions 1–5;
   (ii) What adaptations are necessary to deal with these problems in the ITU Convention and Radio Regulations?

Certain common assumptions and conclusions may be noted in the Replies to the Questionnaire.

Three assumptions underlie all the Replies. First, direct satellite broadcasting will take the form initially, and perhaps for a number of years, of satellite broadcasts to *community* receivers, and it is to this form of broadcasting that the replies to particular questions are largely addressed. Secondly, it is assumed that the primary uses of direct satellite broadcasting will be for education, both cultural and technical, and for information services, and that developing countries are likely to give priority to the former. Thirdly, satellite broadcast programmes having global coverage will be rare, and therefore direct satellite broadcasting will be internationally organized and managed principally through bilateral and regional agreements and arrangements, though ITU, probably in part reorganized, and the Intelsat system will doubtless have important roles.

Certain suggestions and conclusions which are common to the four sets of Replies may be briefly noted.

On the extension of freedom of information by satellite broadcasting, a number of factors stand out: the differing effects on freedom of information of the organization of broadcasting by private enterprise or public authority, and as a monopoly or on a competitive basis; the disinclination of some developing countries to accept extended freedom of information unconditionally; and their dependence for the development of satellite broadcasting on a small number of industrialized countries. It is stressed that proposals for restraints upon the transmission and receipt of information by satellite must be strictly scrutinized.

These considerations lead to the second Question. Here the possible controls by public authority of reception by or retransmission from community receivers is seen as decisive. However, there is agreement that an international code of rules or standards for the programme content of satellite broadcasts would be not only difficult to formulate, but would be of doubtful value and could even lead to undesirable restrictions on freedom of information. Standards might, however, be effectively evolved by informal agreements between broadcasters, perhaps on regional bases, or by the establishment of independent commissions, having the task both of issuing general directives and of conciliation in case of complaints.

By the same reasoning, the protection of the individual in the senses of the third Question is not seen in the Replies as being effectively secured by an international code. Rather it is a problem of defining the jurisdiction of national courts, possibly by the elaboration of conflict-of-law rules, so that the rights of individuals can be enforced with reasonable ease in the vastly extended community created by satellite broadcasting. It is suggested that the courts of the domicile or residence of the plaintiff, at least in a regional system of satellite broadcasting, should have jurisdiction. But it also noted that in such an extended community rectification might often be an adequate substitute for damages.

As regards the protection of broadcasts under the fourth and fifth Questions, whether of the economic interests in the programme contribution, or of the intellectual and artistic integrity of the contribution, or of the broadcast itself, there is a general belief expressed in the Replies that direct satellite broadcasting will not qualitatively change the existing problems, presented by broadcasting or their solutions; and that the 'pirating' of satellite broadcasts, whether by use of 'overspill' or by cable retransmissions, will not create particular new difficulties. Analysis of the principle of 'neighbouring rights' and of existing international agreements tends to confirm this belief.

However, the greatly extended range of satellite broadcasting to developing countries may create quantitative problems: for example, what should be the form and scale of dues payable to copyright owners, and from what funds can payments be made? Suggestions are put forward for the establishment, perhaps under the aegis of Unesco, of an international fund in the interest of developing countries.

The issues raised in the sixth Question have now to be seen in the light of the conclusions of the WARC, which took place after this study had been substantially completed. However, a summary of those conclusions, in so far as they bear on direct satellite broadcasting, has been incorporated in the Introduction.

III REPLY OF THE FRENCH GROUP

ASSUMPTIONS

To assess the problem of direct transmission of television broadcasts from telecommunication satellites, the French group—asked to study the problem—based itself on a certain number of hypotheses. In the course of the work of those carrying out the IBI programme, the most up-to-date information on the technical data and possibilities, as well as the likely achievements in the years to come, have modified, to a certain extent, the original outlook.

Taking account of these modifications, it would seem useful to set out the basic data which the French study took as a starting point.

A. *Technical Data*

It is accepted that direct television broadcasts will be made from synchronous satellites, that is to say, satellites sited on an equatorial orbit 36,000 km from the Earth. It is also assumed that it will be possible to beam satellite broadcasts towards specific regions of the Earth with quite considerable accuracy. The diameter of the minimum zone covered would not exceed 1,000 km.

It is also accepted that, contrary to the hypothesis which was taken as a starting point at the beginning of the work of the national groups, satellite broadcasts will be received in the near future, alongside the present relay system, mainly by community receivers (semi-direct broadcasting).

On this last hypothesis, the programmes will be transmitted to receivers in the true sense of the word principally by cable. It may be foreseen that the receiving sets will often be community ones, especially in the developing countries, and will serve whole groups (villages, schools, religious communities, etc.) but the possibility of private sets in family homes is not to be excluded; this possibility will doubtless shortly be the primary form of reception in the industrialized countries. However this may be, in order to provide television by satellite to the population a country will have to make provision for the installation of hundreds—if not of thousands—of community receivers.

Without altogether disregarding the possibility of receiving programmes directly on individual receivers, the following part of this study considers this possibility as being only a most remote eventuality.

B. *Legal Data*

Several resolutions of the UN General Assembly (in particular resolutions 1802(XVII) of 14 December 1962, and 1962(XVIII) of 13 December 1963) as well as the Outer Space Treaty, which entered into force in 1967, outline at the present time the legal framework of the activities of the states in space, especially as regards telecommunications by satellite. These texts proclaim that the use of outer space should be made for the good of all mankind and that this implies international cooperation. The Outer Space Treaty brings, in particular, into positive law several basic principles which have repercussions on the use of space for broadcasting purposes: freedom of space; application of international law and, especially, of the UN Charter to space activities; state responsibility for national activities in space; and obligations to take into consideration the corresponding interests of the other states. These principles govern all consideration relating to the problem of direct broadcasting.

C. *Political Data*

If there can be no question here of enumerating all the data on the problem of direct broadcasting from a political viewpoint, it is nevertheless worth pointing out some of the principal elements which were taken into account in the preparation of the French study.

The first of these elements is the very appreciable part which direct and semi-direct broadcasting will be able to play in the fight against underdevelopment. Television can make an important contribution to the information and education of the developing peoples; in the poorest countries at the present time there is neither an educational system adequate to provide schooling which alone would allow the people to escape from their lot of poverty, nor mass media which would put them sufficiently in touch with the world-wide civilization characteristic of our times. Direct broadcasting, to the extent that it is a relatively cheap means of reaching these masses and of exercising, due to the power of the photograph, a certain influence on them, allows these two needs to be fulfilled. As against this, Europeans are very much aware of the problem of the survival of civilizations which are rich in authentic cultural values and are confronted with the danger of a levelling out and finally an impoverishment which an abuse of mass media can entail.

The use of direct and semi-direct broadcasting for the purpose of coming to the aid of the developing countries cannot be imagined without international cooperation. However, international cooperation is a prime necessity not only in this respect, but also in general. While a coverage zone of 500–1,000 km may seem of little importance for certain very large countries, it would very often include numerous states in Europe or in Africa, which, consequently, are only able to imagine the problem of such broadcasting in an international context. As for the developing countries, they will not be able, for a long time to come, to benefit from direct and semi-direct broadcasting, no matter what their size—this is the case for India, Pakistan, Indonesia, even Brazil—without cooperation with other states. The French group is deeply convinced that the answer to all the problems which direct broadcasting can raise will have to be sought in international cooperation.

As the organization of broadcasting varies from one country to another—sometimes it belongs to the public sector, sometimes it is conceded to private enterprise, sometimes mixed formulas are adopted—international cooperation should group together either governments or private persons or organizations, or both. In this respect, Intelsat sets in one way an important precedent.

It is in keeping with French thinking to wish that international cooperation, whatever may be its agents, be organized within the framework of appropriate bodies, endowed with the necessary powers to draw up rules, to seek and find answers to concrete problems and, if necessary, to supervise the application of the rules which have been drawn up in common, even to arbitrate on disagreements which might arise.

A last basic factor is the consideration that, at least in the years to come, direct and semi-direct satellite broadcasting will be made, in all probability, principally within the framework of different geographical and political regions of the planet. The simple fact that at no time can all humanity be listening, since at every moment one-third of the world's population is asleep, already curtails the scope of direct broadcasting and makes all idea of universality illusory—except of course for exceptional events. Linguistic and political barriers, the differences between regional civilizations from the point of view of interest, cultural level, moral and social ideas, will doubtless create obstacles which direct broadcasting will only surmount with difficulty in the near future. It follows from this that in assessing the problems set out in the IBI questionnaire emphasis must above all be placed on the regional organization of the broadcasts, at least for the years to come. Yet one must not, of course, completely lose sight of the main principles bearing on the subject, but their study should be undertaken mostly at some later stage of development.

QUESTION 1

What means of promoting and extending freedom of information and of communication does direct satellite broadcasting provide, in particular, for developing countries?

REPLY

The principle of freedom of information means the right of every person not only to disseminate information but also to receive it. This freedom is recognized by a certain number of national and international texts. As our objective is to assess the problem of freedom of information in its relationship with direct broadcasting, that is to say in an essentially international context, we shall only call to mind the latest texts: Universal Declaration of Human Rights, Art. 19; International Covenant on Civil and Political Rights, Art. 19; the recommendation of UN General Assembly, resolution 2448 (XXIII); draft resolution on freedom of information by the Economic and Social Council (res. 756). However, from a legal viewpoint, none of these texts is binding, at the present time. It is only in a regional framework that there are compulsory regulations applicable to a certain number of states at the present time (Art. 10 of the European Convention on Human Rights; and compare the Interamerican Convention on Human Rights).

One may consider that if, in general, the principle of freedom of information and of communication is recognized, in reality its application is threatened. These threats are of two sorts: they may come from the states which in authoritarian systems can have recourse to censorship, but, even in non-authoritarian systems, can often in practice control thought communication because they themselves possess—especially in broadcasting matters—the material supports of this communication. The threats may also come from private people whose commercial interests are not necessarily in the interests of all schools of thought.

In France writers and the legislature are in agreement in considering that, in order to escape this double threat, information must be considered as a genuine public service, and, in particular the means of information which, in countries of relatively restricted size, constitutes the most certain means of reaching the entire national community, that is, broadcasting. Several factors must be taken into account to motivate this point of view: the importance of the influence of modern means of communication on people's behaviour, the educational role which one can assign to this 'mass media', the necessity of ensuring a continuous service whatever the circumstances may be. The author of the most important work to appear in recent times on the subject of broadcasting in France (Debbasch, 1967), gives pride of place in his treatise to this statement: 'the public service character of broadcasting is . . . universally proclaimed'. One again finds this same idea in Section 1 of the Act of 27 June 1964, which states that the ORTF 'ensures a national *public service* of radio and television broadcasting . . . with a view to satisfying the needs of information, culture, education, and entertainment of the public: and see new law of 3 July 1972, Article 1.

Some writers, like French legislation, draw the conclusion from this principle that broadcasting has to be a state monopoly. It is not certain, however, that the idea of 'public service' necessarily leads to this result. What does, nevertheless, appear certain, is that everywhere it assumes some state intervention, if this be only by way of licence to operate, concessions, or regulations for the technical aspects of broadcasts. In other respects, the monopoly, even when held by the state, is not a government prerogative but that of the whole community which may entrust it to an independent body.

When one transposes the problem to the international level, which is that of the present study, one finds that direct broadcasting may have a different result in the industrialized countries, which have a broadcasting network making radio and television available to all the inhabitants, and in the developing countries where such networks do not yet exist.

A. *In the industrialized countries* the introduction of direct broadcasting will mean on the one hand a technical improvement—the possibility of reaching the inhabitants who are most poorly placed from a reception point of view, and eventually the improvement of the quality of the television pictures received thanks to community receivers—on the other hand an increase in the number of broadcasting sources, thus of information. In this connection one may consider, in reply to the question, that direct broadcasting will contribute to 'promoting and extending freedom of information and of communication'.

Nevertheless, it is clear that the increase in broadcasting sources will not fail to raise serious problems for the states which use the monopoly system in broadcasting matters, as these broadcasts run counter to the monopoly which is held either by the state itself, or by independent bodies with its authorization. These problems are well known in several European states, either from the fact that the 'peripheral' stations established outside these national territories beam their commercial broadcasts on them (France), or from the fact of 'pirate' transmitting stations established on board ship on the high seas (off United Kingdom, Holland, Denmark, Norway). In both instances the states concerned have taken steps to protect themselves, which they feel to be their privilege, against intrusions caused by private interests. As regards peripheral stations, France virtually controls them by financial participation. As for the latter, the member states of the Council of Europe have concluded, under the auspices of that organization, an international convention which allows supplies of materials and programmes to be cut off from pirate transmitters (European Agreement for the Prevention of Broadcasts transmitted from stations outside National Territories, signed on 22 January 1965).

It is thus permissible to think that even states with democratic regimes will not necessarily allow, unconditionally, the flooding of their territory with television broadcasts completely outside their control. However, the full impact of the problem will only be felt when broadcasts are capable of being received with total freedom, without any possibility of the territorial authorities intervening, that is to say when receivers will be equipped with individual aerials. The deeper study of this hypothesis, which will only be realized on a wide scale at a later stage of evolution, must be undertaken at a later date. In the system which must be envisaged in the near future whereby the states, by means of community receivers, possess a possibility of intervening, it is to be foreseen that community receivers will be in danger of being under the supervision of local authorities. It would be foolish to think that authoritarian states which, as things stand at present, do not allow certain foreign newspapers into their territory —even though these are well known for their objectivity and their responsible character—would permit community receivers which they had themselves installed on their territory to present any radio or TV programme whatever coming from abroad, without claiming the slightest control over them. In the world today the number of governments claiming the right to keep an eye on the information propagated in their country is going to increase rather than decrease. One may thus ask in what way it is possible, as of now, to find solutions universally applicable to all the industrialized states.

Certainly, the increase in the number of community receivers would seem to result in their escaping, to a certain extent, governmental control, as supervision would become more difficult. Familiarity with authoritarian systems allows one to doubt the soundness of this hope, but, in any case, even if the state does not in practice exercise a strict supervision by this means, the mere existence of a possibility of intervention could always be used by it to bring pressure to bear on the broadcasting organizations.

In fact, it does not seem that the problem can be solved on a universal level. Different regions of the world have different ideas concerning the position of the state with regard to freedom of information and of communication, especially as important political and economic interests may be bound up with broadcasting companies. It is thus within a regional framework that rules should be drawn up which will allow direct broadcasting to make its contribution to the freedom of information.

B. *For the developing countries* the role of direct broadcasting satellites in 'promoting and extending freedom of information and of communication' will probably be even more important. On the one hand, as the Director-General of Unesco emphasized when he presented the results of a world-wide inquiry on information, the existence of adequate means of information is one of the conditions of freedom of information. But a very large proportion of the inhabitants of the developing countries are not able to enjoy this fundamental human right fully because they lack means of information. On the other hand it has been shown, thanks to direct broadcasting, that television is called upon to bring a very important contribution to the education of the inhabitants of the developing countries, as it is known that all progress in these countries is essentially linked to the raising of the cultural level and to general training. However, it is extremely difficult to map out the exact differences between information and education; in Africa, as in Asia or Latin America, one understands information in its widest sense. In short, even a purely educational television system, broadcasting directly to developing countries, will in point of fact increase the flow of information, and in doing this will enlarge the freedom of information and of communication.

In most cases the developing countries will not have the necessary technical and economic potential to operate broadcasting satellites; they must fall back on more advanced countries. However, at present even among the latter there are few able to put satellites in orbit. In any case, the leaders and representatives of countries which would thus receive broadcasts have sometimes expressed their anxiety about finding themselves invaded by programmes from foreign sources, when they themselves have no possibility of reaching their own population.

In these circumstances, it may be asked if the governments of these states would allow the installation of community receivers, which would be completely outside their control, for the purpose of receiving wholly foreign broadcasts. It must also be borne in mind that these underdeveloped states often have an authoritarian system which is naturally not inclined to facilitate and to promote freedom of information and of communication.

It seems to us that technical and economic reasons as well as political reasons necessarily call for international cooperation. It is only within the framework of cooperation between advanced and developing countries that the problems from direct broadcasting can be solved and, in particular, that this new technique may genuinely serve freedom of information.

This cooperation is shown above all in matters concerning access to the satellite. Several UN resolutions as well as the Outer Space Treaty confirm that the exploration and use of space should be made for the good of mankind and for the benefit of all states, no matter what their stage of economic or scientific development, and on equal

Reply of the French Group

terms. The developing countries are thus expressly associated with the benefits which may accrue from the exploitation of space. Technical assistance should be given to them by the UN and its specialized agencies as also by the developed states (GA res. 1348(XIII) of 13 Dec. 1958; 1472(XIV) of 12 Dec. 1959, operative para. A.2; 1721(XVI) of 20 Dec. 1961, op. paras A2–D1; 1962(XVIII), op. paras. 2 & 3; 1963 (XVIII) of 13 Dec. 1963, op. para. II.5; 2130 (XX) of 21 Dec. 1965, op. para. II/11). One should also cite resolution 2448(XXIII) of 19 Dec. 1968 on Freedom of Information which, in para. 5,

Draws the attention of the United Nations bodies and specialized agencies concerned to the continuing need for assistance in the development and improvement of information media in the developing countries in order to enable the latter to share in the benefits flowing from the modern technological revolution and to redress the inequality in this field between the developed and the developing countries . . .'

It follows from this principle that access to telecommunication satellites and, in particular, to direct broadcasting satellites must not only be granted to the developing countries, but that also the developed countries should come to their aid in laying economic and technical foundations for the use of broadcasting satellites. In most cases the developing countries have no premises for making effective use of them: technical and economic assistance should allow them to take part in the new method. This assistance should include not only the siting of the necessary infrastructure, but also the training of technicians and engineers.

It is clear that this action cannot be undertaken without the cooperation of two or more states. A bilateral action is noted elsewhere (p. 11), undertaken by virtue of an agreement made on 18 September 1969 between the United States and India. The size of the Indian subcontinent is such that it is conceivable and feasible to establish a programme for it alone. In other cases, where the developing countries are anxious to receive satellite broadcasts, the rational operation of these broadcasts will necessitate regroupings of interested neighbouring states; thus collaboration should be instituted on a multilateral basis. One may recall in this context a proposal made by Professor Blamont, member of the French delegation to the UN Vienna Conference on the Peaceful Use of Space, in 1968.

This proposal foresaw a multilateral international agency to resolve the problem of use of direct satellite broadcasting. The member states of this agency would be, on the one hand states with confirmed experience in space matters which would put the necessary space facilities at the disposal of the agency and would participate in all the programmes, and on the other hand states which would be 'customers', principally developing countries, which would participate only in programmes of interest to them. If the first category of states were to supply the scientific and technical basis of the undertaking, the 'customer' country would in each case be responsible for the territorial location and the educational part of the experiment. One may also recall in this context a draft resolution presented by the representatives of five Latin American countries to the experts' meeting at Unesco in December 1969 which, to a certain extent, points in the same direction.

The present French legislation concerning the right to broadcasting time can be summarized as follows:—

Outside election times, this right can only be defined by the provisions of Article 4 of the Act of 27 June 1964, according to which 'The Administrative Council [of the ORTF] . . . makes sure that the principal tendencies of thought and the main streams of opinion can express themselves through the intermediary of the Office'. But all the attempts made by the opposition parties to establish this right in their favour have, until now, run up against a refusal by the political majority or the government. At one time, it was attempted to justify the modest place accorded to the opposition parties in the ORTF broadcasts by a rather specious argument: that it consisted of 're-establishing' a balance put in danger by the almost complete takeover of the written press by the opposition (declaration by the Minister of Information, M. Peyrefitte, in the National Assembly session of 30 April 1965). Since then, the government itself has perceived the dangers attached to a monopoly of political information in favour of any group, even if it is in a majority; has not the second ballot in 1965 been attributed to the 'shock effect' brought about by the appearance of the leaders on the little screen which the public discovered at the time of the presidential campaign? Since then, the ORTF has taken care to increase broadcasts in which political opponents confront each other. These broadcasts remain no less at their discretion.

As against this, during the legislative election campaign, a real right to broadcasting time was instituted by an Act of 29 December 1966. However, the practical methods of the workings of this right are open to discussion. Some people deplored that the Supervisory Committee instituted for the presidential election should not be competent for the legislative elections, and the task of safeguarding the equality of the candidates should be confided to the Administrative Council. Above all, the opposition parties have strongly criticized the provision whereby the length of broadcasts would be divided into two equal parts, 'one being given to the groups belonging to the majority, the other to those who do not belong to it'. They have requested the application of the same system as in Great Britain, which means an allotment based on the percentage of *votes* obtained at the preceding election (which, in the particular case of the 1967 French elections, on the basis of the results of the 1962 elections, would have given 36 per cent of broadcasting time to the majority, 64 per cent to the diverse opposition parties). It has also been asserted that in France the parliamentary majorities change, and such a free and easy criterion would be difficult to apply. But the government replied to this latter argument that the 1966 law was made precisely to encourage the regrouping of the French parties around two poles.

The difficulties encountered in the organization of this right to broadcasting time in the case of a country like France make it possible to foretell those which will be encountered in organizing a similar right on a worldwide scale, with the advent of direct television. If this right can be organized in all Western countries at the time of election campaigns, since it is even organized in a country which intervenes as little as does the United States (Art. 315 of the Federal Communications Act), the matter must be much more difficult outside these particular periods.

In conclusion, it seems to us possible to say that direct or semi-direct broadcasting will be able to contribute greatly to the 'promoting and extending of the freedom of information and of communications', particularly in the developing nations. However, on the hypothesis, which serves as a starting point for this study, by which reception will be principally by means of community receivers, the consent, and indeed the cooperation, of the receiving states must be obtained in every case. The best way to succeed in doing this is, without any doubt, to associate all concerned in the broadcasting at some stage of the operations. This association should be effected within the framework of bilateral or, if possible, multilateral cooperation. It is to be expected that in most cases this cooperation will be regional, or at the most interregional, grouping on the one hand the receiving states belonging to a geographical zone, and on the other hand states capable of undertaking the broadcasting. It seems to us that one may say that facilitating this cooperation amounts to 'promoting and extending freedom of information and of communication'.

QUESTION 2

In what ways, institutional or normative, may
   (a) hostile propaganda;
   (b) programmes tending to disturb or erode religious, cultural, or social life;
   (c) advertising;
be limited or prevented in direct satellite broadcasting consistent with the principle of freedom both to impart and receive information.

Before answering this question it may be useful to recall the French system of programme supervision. A distinction must be made between general supervision and particular supervision of certain programmes.

1. *General Supervision of Programmes*

According to Article 4 of the Act of 27 June 1964,

The Administrative Council shall assess the quality and ethical content of the programmes. It shall supervise the objectivity and accuracy of information broadcast by the ORTF. It shall assure that the principal tendencies of thought and main flow of opinions are enabled to express themselves through the intermediary of the Office.

In fact, the Administrative Council is a too top-heavy body to play this role, which calls for continuous action. Further supervision bodies have been set up to carry out this. These are:
— the Programme Committee, operating from a general point of view;
— the Interministerial Liaison Service for Information operating from a more specifically political point of view.

For further modifications see law of 3 July 1972.

(a) *The Programme Committees*

A decree of 22 July 1964 created 'a Radio-Broadcasting Programme Committee and a Television Broadcasting Programme Committee working with the President of the Administrative Committee and the Director-General of the ORTF'.

These Committees have as their mission:

within the limits of their competence and at the request of the President of the Administrative Committee and the Director-General of the ORTF to express opinions on the composition and orientation of programmes as a whole, as well as on the balance between the different types, and to make any suggestions they consider appropriate for promoting the development and quality of the broadcasts. They have the task also of examining scripts and projects for broadcasts submitted to them by the President of the Administrative Committee and the Director-General of the ORTF.

These Committees are formed on a tripartite principle:
— one-third are Public Service representatives.
— one-third are people competent to deal with family and social questions and educational problems.
— one-third are representatives from the world of art, literature, and science (of whom at least one represents the producers, and one the broadcasting directors from the Office).

The powers of these Committees appear to be, in fact, very limited. On the one hand they have a purely consultative role and if they express an unfavourable opinion the only onus placed upon the

Directors of the Office is to ask them for a second deliberation. On the other hand particular programmes are referred to them only when the technical means of the production of the film and its showing on the ORTF makes this possible. In fact, in many instances, censorship is exercised by the management of the ORTF without consulting the Committees; either the broadcast is purely and simply withdrawn from the programme, or a 'White Square' shows the programme is not recommended for sensitive people or children.

Finally, the supervision of the Committees applies to the quality of broadcasts, and it is not within their power to replace the Administrative Council in the ethical or political supervision of broadcasts.

(b) *Political Supervision*

The Act of 31 July 1963 has created 'An Interministerial Liaison Service for Information' (SLII) whose mission is 'to ensure a liaison between different ministerial departments allowing for a continuous coordination of government information and to establish rapid connections between the different means of communication'. In fact, if one believes the caricature drawn by a former Deputy Director of the ORTF, M. Jacques Thibau, it consists of a meeting of ten civil servants every morning who ask each other: '(1) What the television must not speak of; (2) what inaugural and official ceremonies must be widely covered'. The parliamentary opposition quickly made the SLII its favourite target. Due to the protests which it provoked, and following the events of May–June 1968, the government decided to suppress it. An Interministerial Committee for Information was formed, inheriting the functions of coordination of the former SLII but no longer exercising any control over the ORTF. Finally, during the presidential election campaign in May 1969, M. Pompidou promised a further relaxation of political supervision over the ORTF. Actually since the formation of M. Chaban-Delmas's government, the Ministry of Information has been simply suppressed, and the government itself has dealt with questions relating to the ORTF at the level of the Prime Minister himself.

2. *Supervision of Particular Programmes*

Whether it is interpreted as a sign that the general methods previously described have failed, or one wants to see in it the proof of government goodwill in exercising a special supervision in the most important cases, the fact is that recently two series of broadcasts have been the object of such supervision, in fields, moreover, as far apart from each other as possible:

(a) *Commercial Broadcasts*

In face of the arguments which marked the introduction of advertising to the ORTF in 1968, the government felt it necessary to establish special supervision over the substance of these broadcasts. Many adversaries of commercial broadcasts, in fact, denounce the 'hidden persuader' character of these broadcasts which they consider incompatible with a public service of 'information, culture, education, and public entertainment' which is the responsibility of the Office. Therefore, a Technical Commission was formed, presided over by the Director-General of Régie Française de Publicité (RFP), but leaving a large place for state and consumer representatives as well as those from advertising. This Commission has elaborated a real radio and TV commercial broadcasting Code. This Code begins by recalling that the RFP Statutes compel it to respect the fundamental interests of the national economy, and the general responsibilities devolving upon the ORTF (Art. 1). It states the goal sought by the authors of the Code, which is 'to acquaint advertisers as fully as possible with the fundamental rules of morality and accuracy with which they must conform', without, however, there being any obligation on the RFP to accept broadcasts which respect the rules of the Code.

As to the substance of the Code, it is comprehensive, as it deals at the same time with the protection of the French language (the texts must be written in good French—Art. 3); moral, religious, philosophical, and political convictions of the audience (Art. 6); the dignity of the state (Art. 8); the moral, mental, and physical equilibrium of children (Art. 15); public health. Particular rules apply to advertising—promoting medicines while advertising alcoholic drinks and tobacco is forbidden (Arts. 23, 25, 26).

Here the same problem crops up again. These rules are in danger of being totally ineffective with the introduction of direct television unless international rules for commercial broadcasting are instituted. Even if they did not include all the criteria actually observed in the French Code—it is difficult to see how they could take account of all the fundamental interests of the national economy—they appear eminently desirable. The work of harmonizing the national regulations will not be less delicate: for example, the prohibition of the interruption of a programme to insert commercial communiqués. This is allowed in Italy, France, and the UK. Thus, it is not too early to begin to study the problem on a world scale.

(b) *Electoral Propaganda Broadcasts*

With the election of the President of the Republic by universal ballot, by a reform of the Constitution introduced in 1962, the necessity of a standard ruling on electoral propaganda broadcasts appeared. Until then, in fact, these broadcasts were covered by a special ruling drawn up on the occasion of each electoral campaign.

These standard rules were fixed by a decree of 14 March 1964, in fact ahead of the present Statute of the ORTF. The particular interest of this system lies in the formation of a National Commission to Supervise the Electoral Campaign, composed of the highest holders of judicial office. Indeed, at the time of the first campaign for the election of a President (1965), the Supervisory Commission adopted, in matters concerning the televised broadcasts by candidates, a very extensive interpretation of its powers which it defined in a 'directive'. It withheld supervision over 'broadcasts from stations known as peripheral'. Here it confined itself to reminding the candidates and directors of these stations of their responsibilities to the government; in contrast, it exercised supervision very firmly in all matters which concern the ORTF. It forbade, for example, the use of any film on sound production, it compelled candidates to be introduced either in a setting constructed by the ORTF service, or in a setting representing their place of work; it decided that the candidate could be assisted by a journalist of his choice.

In order to enforce the observance of the rules which it has enacted, the Commission did not hesitate to assume in effect a preliminary right of censorship comprising the power to request modifications of projected broadcasts and, in case of refusal by the candidate, to oppose the transmission of the broadcast. In order to exercise this censorship, it has naturally prescribed that all broadcasts are recorded for subsequent transmission. It is hard to see how this system could be applied to direct television broadcasts coming from other countries. Two solutions are thus possible to maintain a balance, which would preserve the equality of the candidates:

—either foreign stations, as is now the case with peripheral stations, spontaneously enforce the rules of equal division of time amongst all the candidates;
—or a strict supervision is exercised over the expenditure of the candidates.

It can, however, only consist of palliative methods which would not replace the real ethical code at present laid down by the National Supervisory Commission. It is therefore very desirable that this ethical code should be incorporated in an international agreement. Be that as it may, the current authoritative system has worked to the complete satisfaction of all the candidates, and all the parties who supported them, during the elections of both 1965 and 1969. However, its application remains confined to the presidential election. Apart from this situation, what constitutes the real rights to broadcasting time for the different political and ideological streams remains to be defined.

REPLY TO QUESTION 2

Considering now the effect of direct television broadcasts on relations between states, we note that on the one hand hostile propaganda, programmes tending to disturb or erode religious, cultural, or social life, and advertising may constitute interventions in the life, political, social, and economic structures of a state, which its government judges harmful. On the other hand, according to the terms of Article VI of the Outer Space Treaty made on 27 January 1967, the states shall bear international responsibility for national activities in outer space, whether such activities are carried on by governmental agencies or by non-governmental entities, and for assuming, on their own responsibility, that the national activities are carried out in conformity with the provisions set forth in the Treaty. This latter—one need hardly recall—declares that space should serve the cause of international cooperation and that it should contribute to the development of mutual understanding and to the strengthening of friendly relations between states and people (preamble, subparagraphs 5 & 6).

It follows from the conjunction of these two factors that direct broadcasting may cause controversies, even disputes between states, on an intergovernmental level, between, on the one hand, states which receive the broadcasts, and on the other, states which participate in the broadcasting. It is not only in accordance with the Outer Space Treaty and with all the principles enunciated by the UN on this subject, but also in the interest of those responsible for broadcasts and in the last analysis for promoting freedom of information itself, that these disputes should be overcome, and if possible eliminated beforehand.

It is indeed advisable not to overlook that in the system envisaged, based on the reception of broadcasts by community receivers, the territorial state, if it considers itself injured in one way or another, always has the possibility of preventing reception by having the receivers disconnected. This hypothesis would be practically unthinkable in a system where there were individual receivers, and would lead to prohibitions in every-day use during war or in countries with tyrannical systems. To a certain extent, broadcasting to community receivers offers greater scope for territorial state intervention than broadcasting by more traditional means.

The methods which may be considered the best for avoiding or overcoming disputes between states, and for thus ensuring freedom of information, differ on certain points as regards the three parts of the question put.

A. *Concerning hostile propaganda*, it is well to remember in the first place that general international law contains a specific rule on one form of propaganda: that which may be considered as an offence, committed in public, against the head or the official representative of a foreign state. The territorial state is bound to ensure the protection of these persons against public offence. In France, the Act on Freedom of the Press of 29 July 1881 (ss. 36, 37, & 48) provides sanctions against the perpetrators of such offences, but nevertheless prosecution in these cases is subject to a request being addressed on behalf of those concerned to the French Minister of Foreign Affairs.

One may also recall another rule of general international law, that which forbids every state to interfere in the internal affairs of other states, but how can the application of this principle be ensured by the traditional rules of international law on broadcasting matters?

Finally, it is fitting to recall the different attempts to forbid hostile propaganda and the most dangerous form of this, war propaganda, in particular: the International Convention Concerning the Use of Broadcasting in the Cause of Peace was signed in Geneva on 23 September 1936, by thirty states, various proposals submitted to the UN General Assembly, and some principles adopted by the latter on the subject (res. 110(II) of 3 November 1947, res. 381(V) of 17 November 1950). However, it does not seem that there is much chance of introducing international regulations on propaganda which would have universal application (V. J. Klein, 1970, pp. 24 & 55). Nevertheless it is necessary to bear in mind Article 20 of the International Covenant on Civil and Political Rights, adopted by the General Assembly on 16 December 1966, and open to signature from that date, which provides as follows:

1. Any propaganda for war shall be prohibited by law.
2. Any advocacy of national, racial, or religious hatred that constitutes incitement to discrimination, hostility or violence shall be prohibited by law.

These provisions, which have not yet come into force, seem however to call for internal legislation, such as exists or is to be drawn up, by the states parties to the Covenant and, for this reason, they are not directly applicable.

Failing the competence to draw up specific regulations which would be universally applicable, the answer to the problem should be sought in international cooperation on a level where this cooperation has most chance of being effectively carried out, i.e. on a regional level. The best means of avoiding disputes arising from broadcasts capable of being interpreted as constituting hostile propaganda is to associate all interested parties in the preparation of these broadcasts or, at least, in the bodies who influence the broadcasts. The regional agencies which were mentioned earlier (Question 1) might, for this purpose, include a special body, made up of three categories of people: those professionally engaged in broadcasting; government representatives; and those well known and independent of either category (each category would hold one-third of the votes) to which disputes could be referred and which would make proposals to those concerned for the settlement of the dispute.

B. *Programmes tending to disturb or erode religious, cultural, or social life.* Although the determination of a case in which a programme 'tends to disturb or erode religious, cultural, or social life' may be extremely delicate, it should be emphasized that this is a real danger which must be faced at the risk of seeing the genuine values of civilization disappear, to be replaced by stereotypes created by political ideologies, by advertising, or by the customs of the consumer society. It does not mean sheltering provinces or whole countries from outside influence, shielding them from contemporary life by converting them into 'reserves' in the way the Indians in the United States live in 'reserves'. It essentially means not destroying certain cultural values of an ethnic or religious community, at a time when that community is still very vulnerable, that is, when the people who are the holders of a civilization are not yet conscious of its true value. It is clear that a country which is much attached to its traditions, such as Switzerland, does not run any risk in receiving any programme whatsoever, whereas it would be a great pity to witness the disappearance of the cultural heritage of populations who in this respect are at a stage comparable to those who, a few centuries ago, exchanged precious stones for glass beads.

Of course, from this point of view again, world-wide regulations would be extremely difficult to draw up and put into force. It would seem once more that the answer should be sought in regional organizations which ought to include bodies grouping representatives of the broadcasters who prepare the broadcasts or take part in them, as well as, on an advisory basis, distinguished people representing various cultures, religions, or philosophical doctrines to be found in the reception zones. These bodies should make rules for the orientation of the programmes and should supervise their implementation. In case of dispute, the conciliation body suggested in point A could be asked to deal with the matter.

C. *Advertising*. As a general rule it is difficult to try to force commercial broadcasts on a country which does not want them—the clampdown of the French government on the peripheral transmitters and even the existence of a European Convention on 'pirate stations' shows this. It will doubtless be the same in the case of broadcasts received by community receivers where the states will keep their full power of intervention. This lends one to believe that within the organization where in the first instance such broadcasting would be carried out—doubtless the regional organization—commercial broadcasts would be subject to rules drawn up beforehand; once the principle of television advertising is allowed, this would constitute some sort of guarantee.

It is not irrelevant in this respect to recall some of the provisions of the ORTF Regulations for radio and television advertising, especially since these regulations bear a strong resemblance to those of the Independent Television Authority.

According to their very terms (Art. 2) these Regulations are for the purpose of 'determining the principles governing radio and television advertising and protecting public interests, by informing the advertisers as fully as possible of the basic rules and of the truthfulness with which they must conform'.

From our point of view, the following provisions deserve special attention:

—Commercial broadcasts shall in so far as possible have an artistic, documentary or educational interest; ensure that the consumer is informed and correspond to an attempt to improve the quality and reduce the price of the goods and services. They shall be free of any vulgarity or trace of bad taste. The texts shall be drafted in good French.
—Advertising shall be honest, truthful, and decent.
—Broadcasts shall not contain any element or any allusion, such as may shock the moral, religious, philosophical or political persuasions of the audience.
—Scenes of violence, scenes which could cause fear shall be excluded, as shall those which might directly or indirectly encourage abuse, imprudence or negligence.
—Commercial broadcasts shall not make an appeal for public charity. In a general manner, appeals for funds shall be forbidden.
—Advertising shall be done in a way which will not abuse the confidence or exploit the lack of experience or knowledge of the consumers.
—Any presentation promoting or suggesting inaccurate qualities of a product or service shall be forbidden.
—The advertisers and their agencies shall be prepared to produce such proof as is necessary to substantiate any description, statement, testimonial, sample, or experiment on which they might be asked questions.
—Advertising shall not contain references, testimonials or other statements from or by an individual, a company, or a particular institution without the permission of those concerned or their legal representatives.
—No broadcast or commercial advertisement shall be capable of causing moral, mental, or physical injury to children; no radio or television advertising system shall take advantage of the impressibility and gift of believing which is a characteristic of children.
—No commercial broadcast shall include games of chance, lotteries, radio or television games.

It could be imagined that in case of dispute over commercial broadcasts the tripartite commission described above (see point A) would be competent.

QUESTION 3

In what ways does the individual require protection in direct satellite broadcasting from:
(a) defamation or false statements in particular by a right of reply and the duty of rectification;
(b) invasion of privacy such as undue publicity of personal life.

REPLY

Protection of the individual against defamation or false statements as well as from invasion of privacy or undue publicity of personal life was one of the aims of the authors of international texts aimed at the international protection of human rights. The Universal Declaration of Human Rights of 10 December 1948 sets out the principle as follows: 'No one shall be subjected to arbitrary interference with his privacy, family, home or correspondence, nor to attacks upon his honour and reputation. Everyone has the right to the protection of the law against such interference or attacks.' The International Covenant on Civil and Political Rights has reproduced this provision word for word at Article 17.

In the European Convention on Human Rights the right of everyone to respect for his private and family life is guaranteed by Article 8, but the protection of the individual against defamation and false statements only follows indirectly from Article 10 which deals with freedom of expression: the exercise of the right of freedom of expression may in accordance with this article be subject to such formalities, conditions, restrictions or penalties as are prescribed by law and are necessary in a democratic society, for the protection of the reputation and rights of others.

Thus, international instruments set forth the principle of protection of the individual against certain attacks. Nevertheless, they leave in all cases the task of defining the conditions of this protection to national legislation.

Concerning direct broadcasting, if it is desirable that the principle of protection of the individual be reaffirmed on the international level—which in any case already results from the instruments for the protection of human rights—it is clear that in given cases this protection will have to be ensured by national courts under the law of the states concerned. It does not seem that direct broadcasting raises problems fundamentally different from those with which we are familiar as of now, at the present stage of radio broadcasting which can be captured abroad or of newspapers which circulate more or less freely in foreign countries. Semi-direct television could, on the contrary, without raising very different problems, modify the extent and acuteness of the current problems by the appreciable increase in the number of people reached by television broadcasts in foreign countries and by the considerably greater impact of television than that of the written press or of ordinary radio.

Hence the main task should be to make the procedure for bringing complaints before courts easier for those who, rightly or wrongly, might feel they are injured by semi-direct television broadcasts. As matters stand in private international law the plaintiffs must apply, as a rule, either to the court which is competent by reason of the domicile of the defendant, or to the court of the place where the offence was committed. In each of the two cases the plaintiffs might be compelled to plead their case in a foreign country, possibly one situated at quite a distance from their homes. It is certain that this situation presents serious disadvantages. It would therefore be desirable that in matters of defamation or false statements or invasion of privacy, the plaintiffs should be entitled to bring their case before the courts of their own domicile. This rule does not seem unfair to the broadcasters because it seems probable that, especially in the system we foresee where there will be regional broadcasters' associations, these organizations will have to have representatives in every country, where their broadcasts are beamed. This new rule of jurisdiction should, of course, be set out in an international convention between the states concerned, in order to modify the currently applicable rules of private international law.

In matters specifically concerning commercial broadcasts, one may recall a rule applicable in France on this subject which is inserted in the ORTF advertising Regulations:

> Commercial broadcasts must not contain any defamatory implication or allusion, or constitute an actionable liability. In particular, these broadcasts should not contain disparaging comparisons with other identifiable brands, products, services, enterprises or organizations. Neither must they try to create or to profit from confusion with other brands, products, services, enterprises, or organizations.

Failing the laying down of a rule of this kind among the advertising conditions, complaints about unfair competition in commercial matters could be likened to those which have been provided for in cases of defamation.

Finally, a 'right of reply' is now recognized in France under the law of 3 July 1972, Article 8. Further it may be supposed that if a court, before which a case of defamation was brought, recognized the justice of the complaint, it would grant the plaintiff the right to insist upon the broadcaster broadcasting a rectification.

One may also conceive that the requests for rectification would be addressed to the tripartite body envisaged earlier which would normally be competent to deal with all controversies, and that this body would decide whether or not a rectification should be made. The first solution—rectification granted by the courts—would obviously call for a modification of the legislation or at least of the judicial precedents of the states concerned. The second solution would seem, to a certain extent, easier, because it could be provided for by a charter organizing the regional broadcasters. The broadcasters would thus set a good example in the cause of the defence of the freedoms and fundamental rights of the individual.

Concerning 'invasion of privacy such as undue publicity of personal life', it deals essentially, it seems, with the right of the individual to his reputation. It goes without saying that this right should be protected every time the individual has been taken by surprise in his private life, quite apart from any participation in public meetings. If the individual who has been filmed as a spectator at a sports event or any similar occasion cannot complain about this, even if he is shown on the screen in an unusual or even ridiculous position, it is necessary that the privacy of the individual be respected. The legal remedies in these cases should be the same as for the case of defamation in the true sense of the term.

QUESTION 4

In what ways, institutional and normative, do authors, performers and producers of broadcast programmes, need special protection for contributions to direct satellite broadcasts in respect of (a) economic interests in the contribution, (b) the intellectual and artistic integrity of the contribution? Do direct satellite broadcasts themselves need special protection?

REPLY

Here again it may be useful to recall the basic principles which are applied in France to protect radio and television broadcasts.

These broadcasts, in so far as they are signals transmitted by Hertzian waves, are protected both by the ORTF monopoly and by what is generally called a 'neighbouring right', that is to say a right which, without being truly a copyright, is nevertheless related to it.

But it must not be overlooked that programmes broadcast by radio or television are made up essentially of creative works and as such they are subject to the right of literary or artistic ownership, or to copyright in the true meaning of the word.

To judge what protection broadcasting enjoys in France, it is therefore appropriate to study on the one hand the protection of broadcasts in themselves, independently of their content, and on the other hand, the protection of programmes, that is to say the radio or television works.

I. THE PROTECTION OF BROADCASTS IN THEMSELVES

The aim of broadcasting is to satisfy a multitude of listeners or television viewers in the privacy of their own homes. Thus there is a temptation for a third party to take over, without making any payment, broadcasts made by a broadcasting organization, and to transmit them for profit to a different and usually paying public. Therefore broadcasting organizations all over the world, whose broadcasts are often very costly, have felt the need to be protected against three types of initiative: the rebroadcasting of programmes, their fixation, and their public use, particularly against payment.

In French domestic law Act 64,621 of 27 June 1964, which gave the ORTF its Statute, appointed this body to inherit the rights of all kinds and the obligations formerly belonging to the Radio-Télévision Française, and which were granted to it by Order 59,273 of 4 February 1959.

Article 4 of this Act provides:

It is forbidden, except with permission granted, subject to the monopoly of the Administration of Posts, Telegraphs, and Telephones, by the Director-General of the Radiodiffusion Française to rebroadcast either by wire or without wire, to record, or to reproduce by any method, all or part of a radio or television broadcast, with a view to broadcasting to the general public either for payment or without payment, subject to identical restrictions to those which are established by Act 57,928 of 11 March 1957 on literary and artistic ownership.

This text thus assigns to the ORTF a neighbouring right to that of copyright. But this right only protects broadcasts against some of the ventures which risk harming the broadcasting body. Thus it is advisable to indicate the other sources which the ORTF can rely on for the protection of its broadcasts.

A. *Protection resulting from Order 59,273*

The neighbouring right recognized in favour of the ORTF allows the latter to oppose the rebroadcasting of its broadcasts. This protection, however, has certain limitations.

Rebroadcasting can occur in two ways: it can either be in the form of a new broadcast, or a distribution of broadcasts by cable.

A new broadcast is the most common and the most dangerous for the original transmitting body. With the help of clandestine relay stations another broadcasting body can thus feed its own programmes without the permission of the original body and at no cost. Let us note that when direct satellite broadcasting comes, relay stations will not even be necessary.

The danger is particularly great in the case of broadcasting of sports events or artistic performances. Fearing that reporting these events would encourage the public to stay away from the places where the events are taking place or that they would harm future productions, the organizers will generally only allow radio or television rebroadcasting (especially when it is 'live') if the broadcast is neither seen nor heard in certain regions or certain countries. If the broadcasting organization cannot guarantee them that there will not be rebroadcasting, they may therefore forbid the rebroadcast or charge much more for it.

In French law, it is true that protection of broadcasts against rebroadcasting without cable hardly raises any problems, given the monopoly enjoyed by the ORTF. Article 4 of the Order made on 4 February 1959, further consolidates this protection, since by virtue of the neighbouring right of the ORTF, the broadcasting without cable of all or part of a broadcast must be authorized by the Director-General of that organization.

It was on these two counts—violation of the monopoly, and attack on the ORTF neighbouring right—that the radio technicians who built rebroadcasting relay stations, without preliminary authorization, on the territory of several Corsican communes were found guilty (Tribunal de Grand Instance Bastia and corr. 29.3.1966; J.C.P.1966, II.14,837).

The rebroadcasting of broadcasts by cable is of interest in regions where physical features prevent satisfactory reception of radio waves, but a clear policy, and its applicability to other regions, has still to be worked out.

If it was, the ORTF could exercise its neighbouring right anew: indeed Article 4 of the 1959 Order makes the rebroadcasting by wire of all or part of a broadcast likewise subject to the authorization of the Director-General.

The protection afforded by Article 4 of the 1959 Order is nevertheless only 'subject to identical limitations to those following the Act made on 11 March 1957, concerning literary and artistic ownership'. These limitations apply in theory to all privileges comprised in the ORTF neighbouring right, but in fact they only really apply in connection with protection against fixation.

B. *The Other Sources of Protection*

The necessity of having complementary protection is obvious, especially in matters concerning the public reception of broadcasts.

As a matter of fact, these can be received in public places—cafés, restaurants, hotels—by means of ordinary receivers, or in specially designed halls which have machines for receiving broadcasts and projecting them on a large screen.

Article 4 of the Order of 4 February 1959 does not speak of public reception of broadcasts. Therefore the ORTF cannot avail itself of its neighbouring right to oppose the direct projection of its broadcasts in cinemas.

Nevertheless, broadcasts are not deprived of all protection against public reception: indeed, a protection is derived from tax provisions which are antecedent to the 1959 Order. The essential part of these regulations follows.

First, the installation and use of any radio or television set makes it obligatory to make a declaration and to pay a licence fee (D.11. Oct. 1958). But the amount of this licence fee or tax is not uniform; it varies according to whether the receiving set is purely for private use or whether it is installed in a place open to the public and, in the latter case, also according to whether entry to the hall for purposes of listening or watching is free or whether payment is demanded. There is some hesitation as to the nature of 'payment' when a café owner uses a radio or television set for the entertainment of his clients; in fact, the Ministry of Finance does not impose the payment of an entertainment tax on café owners. But there is no doubt as to this character of 'payment' when a cinema owner projects a television broadcast on his screen.

Further, the installation and use of a television receiver in a place of entertainment where payment is demanded is subject to the preliminary authorization of the Director-General of the ORTF (Decree of 25 June 1953). This authorization, which is granted after an inquiry, is valid only for the receiving installation described in the application; it is assigned personally to the possessor of this installation and cannot be sold. It is valid for a period of three years, and it can be renewed tacitly.

The corresponding licence fee is made up of a fixed component and a component which is proportional to the returns, on which the tax can vary according to the number of the television performances offered to the public and the dimensions of the screen used. The payment of this licence fee does not give any right to rebroadcast or to reproduce, even partially, broadcasts captured by the authorized receiving installations.

Finally, the installation and use of television receivers in a public place where no payment is asked for entry are subject to identical conditions, with this slight difference, that the licence fee is solely a fixed charge and its amount is lower than in the preceding case.

C. *International Agreements*

Two international agreements, which have been signed by France, are of interest to our subject:

The Rome Convention, for the protection of performers, record producers, and broadcasting organizations, was signed on 26 October 1961, and came into force on 18 May 1964. Even though its ratification has encountered serious obstacles, it seems useful to summarize its principal provisions, obviously only retaining those which concern protection of broadcasts.

The European Agreement on the Protection of Television Broadcasts signed on 22 June 1960, and at present in force in the following countries: Belgium, Cyprus, Denmark, France, German Federal Republic, Norway, Sweden, UK. It has been the subject of an additional Protocol signed on 22 January 1965. (The decree proclaiming this additional Protocol has just been published in France—Decree 70,300 of 2 April 1970, which appeared in the *Journal officiel* of 9 April, p. 3365.)

Although stating certain identical principles, the two Conventions are quite different, both in their scope and in their contents. Thus each should be examined in turn.

1. *The Rome Convention*

Having explained the general principles governing broadcast protection, we will see that the Rome Convention ensures broadcasting organizations a minimum protection which can nevertheless be restricted by the contracting countries.

(a) *General principles*. The Convention has adopted the principle of national treatment, making it binding on member states to comply with a conventional minimum. However, a material reciprocity is insisted upon in certain cases. As a rule, with regard to broadcasts, each contracting state undertakes to grant national treatment if the head office of the broadcasting organization is situated in another contracting state, or if the emission is broadcast by a transmitter sited on the territory of such a state. However, the contracting parties can reserve to themselves the right to protect broadcasts only if the two qualifying criteria are combined, which assumes at the same time that both the head office of the broadcasting organization is situated in another contracting state and the broadcast is transmitted by a transmitter sited on the territory of this same state (Art. 6.)

'Broadcast' in the Convention is understood as meaning: 'Sound transmissions, television transmissions or other types of transmission intended for reception by the public' (Art. 3(f)). This embraces all radio and television broadcasts either 'live' or recorded, except those broadcasts intended for a single individual or a well-defined group such as aircraft in flight or taxis.

The Convention provides special protection for broadcasting organizations which do not break the law of copyright (Art. 1).

The minimum period of time in which broadcasts are protected is twenty years, to be counted from the last day of the year in which the broadcast has taken place (Art. 14(c)).

(b) *The substance of the minimal protection.* The Rome Convention ensures a more complete protection for broadcasts than the Order of 4 February 1959, since it envisages four kinds of initiative (Art. 13). We will consider only two of them here, the protection against rebroadcasting, and the protection against public reception. As to the first, broadcasting organizations have the right to authorize or to forbid the rebroadcasting without cable of their transmissions, but only when this rebroadcasting is relayed simultaneously, i.e. at the same time as the original broadcast. The Convention does not protect relaying at a later date so that, in theory, a contracting country could without authorization rebroadcast an emission produced in another contracting country. However, in point of fact, there is little danger as most internal legislation has regulations on this question.

As for the protection against public reception, broadcasting organizations have a recognized right to authorize or forbid 'the communication of their televised broadcasts to the public when it is done in places which are open to the public, on payment of an entrance fee'.

The Convention in this way wished to penalize wrongful appropriation of broadcasts by the proprietors of public establishments and to make the rebroadcasting of artistic performances and sports events possible. As we have seen, in order to do this broadcasting organizations must be able to forbid public reception of their broadcasts within a determined area, or make them liable to supplementary payment.

It will be observed that protection against public reception only concerns television broadcasts, and putting it into practice presupposes two conditions being fulfilled. It is necessary in effect:
— that the reception of the broadcast takes place in a place open to the public, which covers all entertainment businesses, especially cinemas;
— that an entrance fee is collected, which excludes, it seems, cafés, hotels, and restaurants even though the showing of broadcasts in these establishments is not disinterested.

However, the actual working terms of the right granted to broadcasting organizations are rather uncertain because the Convention leaves the task of defining them to the domestic legislation of the country where protection is requested (Art. 13). Consequently, the contracting states seem to have wide scope for saying what must be understood by 'places open to the public' and by 'entrance fee'. Moreover, they can refuse to acknowledge protection against public reception (see below).

(c) *Limitations on the minimal protection.* Even though in theory it is minimal, the protection of broadcasts which arises from the above provisions can be curtailed owing to a set of options or reservations which are provided by the Convention.

A state may declare that it will not grant broadcasting organizations a right of public reception for television broadcasts (Art. 16(b)).

In such a case, a reciprocity clause operates: if a state declares that it will not protect, on its territory, foreign television broadcasts against public reception, the other contracting states can reciprocally refuse the conventional protection to broadcasting organizations which have their head office on the territory of the first state.

The reservation relating to public reception can be made not only at the time of deposit of the instruments of ratification, acceptance, or membership but afterwards at any time; in this event it will only come into force six months after it has been lodged.

All contracting states have the option of providing in their internal legislation for exceptions to the protection guaranteed by the Convention in the cases which are set out in Article 16(1). The most important of these exceptions concerns private use.

This exception is quite usual since it is allowed by all internal legislation on copyright. But the idea of 'private use' is not understood everywhere in the same way. In this regard, French law is particularly restrictive: indeed, it only tolerates 'private performances (with no payment involved) executed *exclusively within the family circle* and reproductions strictly reserved for the private use of the copyist' (Act of 11 March 1957, Art. 41 (1) & (2)). If the same meaning of private use is used for the interpretation of the Rome Convention, it must be decided that a broadcast in a factory, cultural centre, or hospital is not of a private nature, and consequently cannot benefit under the first exception provided for by Article 15(1) of the Convention.

In addition to the various exceptions, each contracting state is allowed to provide in its internal legislation for 'similar limitations' to those which are provided in this legislation, concerning the copyright protection of literary and artistic works. However, it is specified that 'compulsory licences can be instituted only to the extent that they are compatible with the provisions of the present Convention' (Art. 15(2)).

Thus the possibility given to contracting states of further limiting the conventional protection is of little importance in French law. Indeed, the rare restrictions which the latter makes on copyright are already present amongst the exceptions provided by Article 15(1) of the Convention; besides French law completely disregards legal licences in the subject of literary and artistic ownership.

2. *The European Agreement on the Protection of Television Broadcasts*
This Agreement, whose application is limited to Western European countries, serves to protect television broadcasts with a view to promoting the exchange of televised programmes. It has, thereby, a more restricted scope than the Rome Convention. The additional Protocol of 22 January 1965, lays down, however, that from 1 January 1975 no state can remain or become a party to the Agreement without being also a party to the Rome Convention of 26 October 1961 (Art. 3(2) of the Protocol modifying Art. 13 of the Agreement).

To enable us to establish a comparison between the European Agreement and the Rome Convention, we will follow the same order as before in distinguishing between the general principles, the conventional protection, and the limitations which can be brought to bear on it by the contracting states.

(a) *General Principles.* Like the Rome Convention, the European Agreement lays down the principle of national payment with an obligation on the member states to respect a conventional minimum and, in certain instances, recourse to material reciprocity.

Each contracting state undertakes, as a rule, to protect broadcasting organizations which are established within the territory of a party to the Agreement in accordance with its law, and those which make their broadcasts within such a territory. They have, however, the option of reserving the application of the Agreement to broadcasting organizations which comply with both biding criteria, that is those which are established within the territory of a party in accordance with its laws and which perform their broadcasts within this same territory (Art. 3(1)f).

The Agreement has been drawn up in order to apply to national as well as international situations. Each state can, however, exclude from protection broadcasts effected by organizations established within its territory in accordance with its law, or broadcasts effected within this territory, when these broadcasts will enjoy a protection under their domestic law (Art. 3(1)e).

The protection laid down by the European Agreement only covers television broadcasts, but the latter benefit from it whether they be made 'live' or are filmed and in their sound element as well as in their visual element. The protection does not apply, however, to the sound element when it is broadcast separately by a broadcasting transmitter (Art. 5).

The terms of the Agreement do not affect copyright, or the rights of creative artists, actors or performers, film producers or record producers, and organizers of performances. They also leave untouched the protection of the broadcasts based on other sources such as unfair competition (Art. 6).

The minimum duration of protection granted to television broadcasts which was, at the beginning, fixed at ten years (to be counted from the end of the civil year during which the first broadcast has been made within the territory of a party to the Agreement) has been extended to twenty years by an additional Protocol on 22 January 1965 (Art 1, modifying Art. 2(1)a of the Agreement).

(b) *The substance of the minimal protection.* The European Agreement ensures a more complete protection for broadcasting organizations than that of the Rome Convention.

Broadcasts are protected against all rebroadcasting, by wire or without wire, simultaneous or deferred (Art. 1(a) & b).

Broadcasts are protected against any communication to the public by any instrument for transmitting signs, sounds or images (Art. 1(1)c) and even by means of fixations or reproductions unless the organization holding the right to them has authorized the sale of these fixations or reproductions to the public (Art. 1(1)e).

Supplementary protection is provided for in the case where the fixation of a broadcast or its reproduction should be made either within a territory to which the Agreement is not applicable, or in a country which is a party to the Agreement, which case will be discussed later: if these fixations or reproductions are imported within the territory of a party to the Agreement, where they would be illicit without the authorization of the organization holding the right to them, they could be seized there (Art. 4).

(c) *Limitations to the minimal protection.* In the same manner as the Rome Convention, the European Agreement provides for several possibilities of limitations and reservations in favour of the contracting states, which enable them to curtail, within their territory, the protection guaranteed as a minimum conventionally.

The contracting states can decide in matters of rebroadcasting:
— not to protect broadcasting organizations established or broadcasting within their territory against broadcasting by wire;
— to limit the exercise of this protection, in the case of organizations established or broadcasting from within the territory of another party to the Agreement, to a percentage of their broadcasts which cannot be below 50 per cent of the average weekly duration of the broadcasts of each of these organizations (Art. 3(1), modified by Art. 2(1) of the additional Protocol).

The contracting states have a further possibility of limiting the protection: when at the time that a report of a news event is made, there is a rebroadcast, fixation or reproduction of the fixation, distribution by, or communication to the public of, short excerpts of a broadcast constituting itself all or part of this event (Art. 3(2)a). This exception, which is very limited, concerns notably the case where a speech is delivered by a head of state in the television studios or in his own home in the presence only of technicians from the broadcasting organization. It is justified by the fact that no other broadcasting organization has the possibility of sending their journalists or cameramen to that place in order to make the broadcast and that the latter constitutes, in itself, a news event of which the public should be able to be informed.

Finally, the contracting parties have the opportunity, in those matters which concern their territory, of appointing a body to which complaint can be made that the right either of distribution by cable, or of communication to the public has been arbitrarily refused or granted on exorbitant terms by the broadcasting organization which holds this right (Art. 2(3) of the Protocol modifying Art. 3(3) of the Agreement).

Such are the rules which, on a national and international level, ensure the protection of radio and television broadcasts. But this protection, it is important to remember, concerns the broadcasts themselves and not the substance of these broadcasts, that is to say the programmes which are broadcast. These programmes are frequently made up of radio and TV works which, in their turn, profit from a protection.

II. PROTECTION OF RADIO AND TV WORKS

In addition to the specific protection which broadcasting organizations enjoy for their broadcasts, quite often there is also the protection relating to creative works. It is not unusual, indeed, for radio or television programmes to have a bearing on such works. Sometimes it is a question of pre-existing works which are already fixed on a material support (cinema films, musical compositions recorded on records) or performed in public (plays, concerts, music hall shows) for which the transmitting organization merely carries out the broadcasting. It may also concern sequences specially made by the staff of that organization for the purposes of broadcasting.

The two hypotheses must be carefully distinguished. In the first case, a literary or artistic work is incorporated into the programme, but it has been created or performed outside the transmitting organization. The latter confines itself to carrying out the technical operations which do not deserve to be protected by copyright. This right naturally comes into the business relations between the broadcasting organization and the author of the work which is transmitted, but the examination of these relations seems to us to go beyond the scope of this study.

In the second hypothesis the question arises of knowing if the broadcasting organizations can claim, in addition to its neighbouring right over its broadcasts, a copyright for radio and TV 'works' created within itself by its own staff and of which it is in a certain way the producer. To answer this question, it is advisable to examine the conditions under which copyright is protected. After that, we will examine the rights relating to radio and TV works.

If copyright is not subject, in France, to any formality, it assumes, at all events, the existence of a work whose nature and characteristics justify its being protected by virtue of literary and artistic ownership. On the other hand, copyright protection is only granted to the person or persons who qualify as the author.

A. *Rights attaching to the work*

Radio and TV works do not figure in the enumeration of protected works given in Article 3 of the Act of 11 March 1957. They are protected nevertheless, because this enumeration is not limitative and it is always necessary to go back to the main principle which rules French law in this matter. The right of literary or artistic ownership applies to any creation of an original form (the originality can, moreover, only be relative) whatever the category, merit, or intended purpose (Act of 11 March 1957, Art. 2).

Further, the comparison established in Article 18 of the Act between radio and TV work on one hand, and cinematographic works on the other, suggests that nearly all sequences carried out by the broadcasting company's agents have a claim to copyright. Indeed, according to French law, documentary films are treated in the same way as works of the mind, and simple photographs are protected like creative works when they show artistic or documentary characteristics (1957 Act, Art. 3). The same protection is consequently applicable not only to cultural or entertainment films ('dramatic' television films, radio or television serials) but to audio-visual commentaries or to the television news magazine. The originality of the commentary, the choice and presentation of the news items and images are enough to make such a sequence into a work which is entitled to copyright.

Moreover, it is necessary to be the author in order to benefit from this protection.

B. *Rights attaching to the author*

In French law, copyright can only belong, at least in the capacity of originator, to the author himself, that is to say he who, alone or with others, has made an intellectual creative work. This principle is applicable to radio and television works as is shown by Article 18, para. 1 of the Act of 11 March 1957, under which the copyright holder must 'Have the qualification of being the author of a radio or television work the person or persons who effect the intellectual creation of such a work'. The formula is practically the same as that used in connection with cinematographic works (Art. 14, para. 1).

But who are the persons who 'ensure the intellectual creation' of the work and who are, thereby, invested with the qualification as author? (a) Usually the radio or television work will be a work of collaboration and the qualification as co-author will then belong to various persons who have worked together for its elaboration (see already before the 1957 Act, Cour de Paris—proclamation of 7 July 1955, Ann Prop. Ind. 1956, 210; R.I.D.A. 1956 CXII 167; *Rev. trim. dr. com.* 1957, 123). As in the case of the cinema it can be a question of the author of the scenario or the adaptation, it can be a question of the author of a spoken text, of musical compositions with or without words especially written for the work, of the producer, even the machine operator or cameraman. A certain number of observations however deserve to be made.

—It will be noticed first of all that Article 18 of the 1957 Act does not refer to Article 14, para. 2, according to which certain persons are presumed, unless there is proof to the contrary, to be co-authors of a cinematographic work which has been effected by collaboration. In radio or television matters it is therefore necessary, in the absence of any legal presumption, to identify in every case those who have actually taken part in the intellectual creation of the work, both in its sound and in its visual elements.

—It will also be noted that the position of machine operators and cameramen can be very different according to the circumstances, in documentary films, in news sequences or in those bearing on a sports event, these technicians play a leading part since the quality of the work largely depends on their agility, presence of mind, and choice of filming angle: the cameramen and machine operators can generally be considered as co-authors of the same rank as the commentator. On the other hand, their part can prove less important in the case of creative works; in such a case, the cameraman will undoubtedly have the qualifications of author of his film shots, but he cannot be regarded as co-author of the entire work.

—Finally it should be pointed out that the actors who render or perform a radio or television work cannot under any circumstances figure as co-authors of these works. They can at the very most avail themselves of a neighbouring right for their performance (Judgment given in the Furtwangler case by the Supreme Court of Appeal, 4 January 1964: D. 1964, 321. J.C.P. 1964, II.13,172) even though this right may not be recognized by French law.

(b) As the result of collaboration between several authors, a radio or television work will sometimes be a composite work, that is to say, a new work which incorporates a pre-existing work with the authorization of the author of the latter (if it is not yet public property) but without his or her collaboration.

Article 18, para. 2, in this case, extends the solution originally provided for by Article 14, para. 3, for cinematographic matters to radio and television works. When a work is drawn from an already existing work or a scenario which is still protected, the authors of the original work are considered on an equal footing with the authors of the new work. They are considered as co-authors of the latter, even though they have not participated in its production.

Hence, for example, when a radio serial or television serial is drawn from a novel, the writer is co-author of the radio or television work, even if he has not taken part personally in the work of adaptation.
(c) Article 18 excludes any possibility of recognizing a corporation or broadcasting organization as an author; which means that the ORTF will never be either author or even co-author of radio or television works which it has produced and it cannot be vested with copyright on the score of being the originator.

C. *The Rights of Co-Authors*

Like any creator of a literary or artistic work, the co-author of a radio or television work enjoys privileges of an intellectual and ethical nature and patrimonial privileges.

Article 18, para. 2, refers back on this point, also to the provision relating to cinematographic works (Art. 15). It is thus established that every co-author has the option of using his personal contribution in a different kind of way, on condition that the common work is not prejudiced and subject, besides, to stipulations to the contrary provided for in the agreement with the other co-authors or the broadcasting organization.

But what is of particular interest to us is knowing what are the rights, taken as a whole, of co-authors of radio and television works. In this respect, it is necessary to distinguish between privileges of an ethical nature and privileges of a patrimonial nature (or monopoly of exploitation).
(a) First, the moral right of the author comprises several attributes: the right of disclosure (with its corollary, the right of non-disclosure), the right to claim authorship, the right to respect, the right to withdraw or repent, and it is considered in French law as being attached to the person of the author, and therefore inalienable.

In order to prevent the exercise of this right from bringing the production of a radio or television work to a standstill, the legislator has extended a provision, which is provided for cinematographic works, to this category of works; Article 18, para. 2, refers back indeed to Article 15, para. 1, which prescribes:

If one of the co-authors refuses to complete his contribution to the work (cinematographic) or finds it impossible to complete it for reasons outside his control, he cannot oppose the use of the part of this contribution which has already been completed, for the purpose of completing the work. He will be considered as author of this contribution and enjoy the rights appertaining to it.

But once the work has been finished and broadcast, common law takes over again. The privileges making up the moral basis being unassignable, the co-authors of a radio or television work retain them even if they have made their rights of exploitation over to the broadcasting organization and they alone can exercise them, either against this organization, or against a third party who might attack their moral interests.

For this reason every author can insist on his name and description being mentioned (see, e.g., with regard to the television serial 'Jacques le Croquant', the judgment of the Tribunal de Grand Instance in Paris on 13 December 1969, given against the ORTF in favour of Mlle O'Glor, co-adaptor and co-script writer of the work; *Le Monde*, 16 Dec. 1969; this judgment has just been upheld by the Cour de Paris, ibid., 5 June 1970).

As for the right to withdraw or repent, it seems that its exercise should be subject to the unanimous agreement of all the collaborators or failing this, to court arbitration, in accordance with all general principles governing collaboration (Act 11 March 1957, Art. 10).

(b) The exploitation monopoly covers both the reproduction right and the right of presentation or public performance, both of which are protected under criminal law against infringement of copyright. Unlike the moral right, this monopoly can be partially or wholly assigned.

In matters of a cinematographic nature, the Act of 11 March 1957, establishes a presumption according to which the contract binding the co-authors to the producer implies, without a clause to the contrary, the assignation of the rights of exploitation in favour of the producer (Art. 17, para. 3). There is no such presumption for radio or television works as Article 18 relating to this category of works does not refer back to Article 17.

But radio or television exploitation rights being property rights can be the subject-matter of controls. Even though it is not the author the broadcasting organization can thus get these rights made over to itself.

D. *The Rights of the Broadcasting Organization*

By virtue of a Decree of 4 February 1960, 'French broadcasting is free to dispose as it wills of services carried out within the framework of its office'.

Actually this choice of words is unfortunate, since as we have said, the ORTF as a corporation can neither be sole author or even co-author of the radio or television works it broadcasts. Even if these works have been produced by its agents, the ORTF therefore is never vested with the original copyright and under no circumstances can it avail itself of the moral right. As against this what does happen is that the ORTF acquires, by means of a transfer implicitly or expressly included in the statute of its agents, the property rights necessary for the exploitation of these works. Therefore it is only as assignee that the broadcasting organization enjoys the protection attached to the patrimonial aspect of copyright and that it can, if necessary, take action against infringement of copyright where the broadcast programmes are unlawfully used.

However, the exact extent of the transfer in favour of the broadcasting organization remains to be defined.

(a) In certain respects, the broadcasting organization's rights are of more consequence than those which are usually recognized in the case of assignees of patrimonial aspects of copyright. Normally, indeed, contracts which have a bearing on the use of copyright are restrictive (1957, Art. 31) and the assignee is only vested with privileges which the author has expressly relinquished in his favour. Thus Article 45, para. 1 of the Act of 11 March 1957, slightly waives these principles on behalf of the broadcasting organization: 'Unless there is a stipulation to the contrary, the authorization to broadcast a work or to communicate it publicly by any other means of transmitting signs, sounds or images than by wire, covers all communications made by the organization which is the recipient of the assignment.'

By thus including, in the authorization to broadcast, relay stations operated by the recipient of the initial permission, this text makes an exception to the principle that all new communication to the public is subject to preliminary authorization by the author or co-authors; each radio or television relay is indeed a communication separate from the others—in spite of being simultaneous—and theoretically all relays should have been subject to right of performance. Article 45 wanted to avoid this complication in the case of rebroadcasting other than by wire, and, it seems, in the case of broadcasting by wire. It will be noted, however, that the derogation allowed by this article is only valid for relays made by organizations which are assignees of the broadcasting right, to the exclusion of rebroadcasts made by other organizations, and that co-authors can set it aside by an express stipulation.

Paragraph 2 recalled that 'the authorization to broadcast does not imply the authorization to record a broadcast work by means of instruments capable of fixation of sounds or images'.

Paragraph 3 of the same articles adds:

However, exceptionally by reason of the national interest they represent, or their documentary character, certain recordings can be authorized. The conditions and terms of their production and utilization will be settled by the parties or, failing an agreement, by a decision signed jointly by the Minister of Fine Arts and the Minister of Information. These recordings shall be preserved in the official archives.

Even though the term 'ephemeral' may not be used, this provision is patterned on Article 11(a) of the Berne Convention relating to ephemeral recordings. The 1957 Act wished to facilitate radio broadcasting of performance in which artists could not take part at the desired time. The recording of the performance allows of a time-lag between the performance of the work and the broadcast, which must not exceed 24 hours. This recording however can be preserved by reason of its documentary character or national interest after it has been used ephemerally. Thus interpreted, in the light of Article 11(a) of the Berne Convention, Article 45, para 3, has a rather limited scope. Nevertheless, it forms a new derogation to Common Law since it can happen that the organization which is assignee of the right solely to broadcast can be authorized to fix the work on a material support.

(b) But apart from these two provisions, the broadcasting organization does not profit from any favours and the assignment which it has been granted expressly or tacitly is subject to restrictive interpretation. This organization is not allowed to make use of a work in any other way than for the radio or television use for which it was intended. Thus, for example, the ORTF cannot in the absence of a stipulation expressly authorizing it, negotiate with distributors so that television films it has made should be shown in public halls, because a showing on a cinema screen differs from a broadcast intended for the television screen; further, it cannot by itself institute exchanges of television films even though such exchanges would lead to important saving of money and thereby reduce the expenses of programme production; the express consent of the co-authors is necessary.

To encourage television programme exchanges the European Arrangement on programme exchange by means of television films of 15 December 1958, however, allows of an exception to this principle (Art. 1). Broadcasting organizations of member states can institute exchanges of television films which they have produced without the preliminary agreement of the co-authors. This exception is only valid however for use made of them for purposes of television and is applicable only to contributions specially produced for the broadcasting organization.

In addition, Article 6 of the Arrangement expressly reserves:
—the moral right recognized in relation to film;

—the copyright in literary, dramatic, or artistic works from which television films are derived;
—the copyright in a musical work, with or without words, accompanying a television film;
—the copyright in films other than television films;
—the copyright in the use of television films otherwise than on television.

Finally, in the event of France ratifying the Rome Convention, it should be pointed out that the principles governing relations between the ORTF and the co-authors of radio or television works produced by its courtesy will not be modified. Doubtless Article 13 of the Convention recognizes the right of the broadcasting organization to authorize the rebroadcasting of its broadcasts, their fixation, and their communication to the public (see above). But these authorizations, which will be given by the broadcasting organization by virtue of its neighbouring right over its broadcasts, will not affect the copyright to radio or television works; as assignee solely of the right to broadcast, the ORTF thus will only be able to authorize other types of use provided for by Article 13 if the co-authors express agreement.

III. PROTECTION OF DIRECT SATELLITE BROADCASTING

If we now consider the question of the protection of authors, performers, and producers of broadcast programmes for contributions to direct-satellite broadcasting; we may admit as a first principle that semi-direct or even direct broadcasting will not modify qualitatively the problems which all broadcasting raises in general in the field of intellectual rights. The changes that may be predicted will be more of a quantitative nature: the broadcasts will perhaps be the same—at least in principle—but the number of television viewers will be greatly increased. We leave aside here examination of the situation in the developing countries; the problems raised by the latter will be assessed under Question 5.

(a) *The economic interests of authors, performers, and producers as well as those of producers of records.* These persons are naturally inclined to demand increased payment when the programmes in which they take part are seen by a large audience. This question, however, should be discussed first of all between the authors, performers, and producers concerned on the one side and the broadcasters on the other; it seems there are certain precedents in this subject, notably in the United States where the fee for licences permitting the broadcasting station to use copyrighted music is fixed on the basis of the size and power of the station, the size of its audience, and so on. This principle can be considered as fully applicable in equity to commercial broadcasts: to the extent that direct broadcasting increases the audience of commercial broadcasts, it would also increase the profits of the broadcasters. This increase would be used to discharge supplementary payment due to authors, performers, and producers of broadcast programmes, as also to producers of records.

The only question which might arise on the intergovernmental level would be to know if, should the case arise, the receiving state should assume responsibility for the difference in the amount of royalties due for the broadcasting of programmes within its territory. It is clear that if the state in question has not requested the setting up of this service, it is difficult to ask that state for any participation whatever, even if it accepts the situation by installing or allowing community receivers to be installed within its territory. It is different when the state concerned has requested that a direct television network serve the territory or that it may make use of it. However, this hypothesis will most often be that of developing countries and so it will be examined under Question 5. In the other cases it may be foreseen that the state wishing to benefit from a direct or semi-direct television system would be associated with the broadcasters, and should for that reason participate in the expenses of the undertaking.

(b) *The intellectual and artistic integrity of the contribution.* In the same was as any creator of a literary or artistic work, the author of a radio or television work enjoys privileges of an intellectual and ethical nature. Under French legislation the moral right of the author comprises several attributes: the right of disclosure (with its corollary, the right of non-disclosure), the right to claim authorship, the right to respect, the right to withdraw or to apologize: these are considered in French law as being attached to the person of the author, and therefore inalienable. For this reason every author can insist on his name and description being mentioned. There are also several international conventions in this field: the Rome Convention, the European Agreement, the Berne Convention, and the Universal Copyright Convention.

It would be advisable to recognize in the field of direct or semi-direct broadcasting that authors, performers, and producers of broadcasts should be able to command the respect of their moral interests, notably the right to have their authorship of the work acknowledged. Moreover, it seems fair to ask both the broadcasters and the states receiving programmes directly transmitted by satellites to undertake to do nothing to alter the artistic or intellectual integrity of the contributions of the authors, performers, and producers, without the consent of the copyright holder. In particular, one should refrain, unless with the consent of those concerned, from all interference with the sound element, visual superimposition or insertion of advertising matter.

(c) *Protection of broadcast.* Until now, protection of radio transmission broadcasts is ensured on the national and international level by a 'neighbouring right' assimilated to copyright which is recognized in favour of the broadcasting organization. The regulation of this right is based on the idea that the broadcasting organization must have the power to authorize or forbid all public utilization of the broadcast. In particular in France Article 4 of the Act of 27 June 1964, which gave the ORTF its Statute provides:

It is forbidden, except with permission granted ... by the Director General of the Radiodiffusion Française, to rebroadcast either by wire or without wire, to record, or to reproduce by any method, all or part of a radio or television broadcast, with a view to broadcasting to the general public, either for payment or without payment, subject to identical restrictions to those which are established by the Act ... of 11 March 1957, on literary and artistic ownership.

This text, completed by a certain number of others, thus assigns to the ORTF a neighbouring right to that of copyright. It should be added that two international treaties on this subject have also been concluded: the Rome Convention, of which the guiding principle is the application of the national system to programmes received from other states and which affords protection for a twenty-year period; the European Agreement which was completed by the Protocol of 22 January 1965. Both these texts likewise lay down the principle of national treatment, with an obligation on the states who are party to them to observe a conventional minimum, and they provide on some points for recourse to material reciprocity. It may be asked to what extent these rules can be applied to a direct broadcasting system. In fact, countries which broadcast by direct satellite transmission necessarily accept that their broadcasts will be received by countries covered by these satellites. In these circumstances there would be, generally speaking, no purpose in giving them a right of authorization or interdiction.

In matters which more especially concern semi-direct television, i.e. television received by community receivers, it is clear that the retransmission of broadcasts by cable is part of the system and cannot be criticized. It might be different for microwave retransmission. Without wishing to go as far as forbidding any use of a broadcast thus received—possibly at a later date—it would be advisable to forbid the rebroadcasting of whole programmes or that of performances for which the broadcasting organization itself has sometimes had to pay large sums of money (sports and variety programmes, plays or opera shown 'live'). In this field again, direct arrangements between those concerned—regional and national broadcasting organizations—will doubtless solve the overall problem.

QUESTION 5
What problems do the issues in Question 4 create for developing countries in the reception of direct satellite broadcasts or access to satellites?

REPLY

The main problem raised by direct or semi-direct broadcasting to the developing countries in copyright matters is that of reconciling the needs of these countries with respect for the rights and interests of all kinds of authors, performers, and broadcasting producers, as also those of record producers.

Concerning the moral interests, no special problem seems to arise; the solutions considered earlier remain valid. But the same does not hold true for the financial problems. On the one hand it is hardly possible to ask the poor countries, for whom the broadcasts are intended, to pay royalty fees; on the other hand, it would be rather unfair to ask those concerned—authors, performers, and broadcasting producers—to carry alone the burden of a form of aid to developing countries.

As regards the broadcasts themselves, there is a distinction to be made between broadcasts specially made for broadcasting to developing countries (principally educational and cultural broadcasts, but also perhaps original artistic broadcasts), and broadcasts which the stations only rebroadcast 'live' or those which the stations would broadcast at a later date to the developing countries.

In the first case the broadcasters will, of course, be responsible for the royalties. It would be conceivable—and sometimes even desirable—that in order to meet these expenses they would have support from the governments which give aid to the developing countries; moreover it is obvious that these broadcasts are a very effective form of such aid.

For broadcasts not originally intended for the developing countries, it would be important that at all events the principle of copyright be protected. It is obvious that producers might agree to special fees, sometimes even nominal ones, for the rebroadcasting of programmes which have already been shown on television. It would also, nevertheless, be conceivable to think of the setting up of an international fund, fed by contributions from the industrialized states, for the purpose of remunerating authors, producers, etc. for broadcasts to the developing countries.

It should specifically be noted that as legislation stands today in states like France, there is no legal means of persuading—and even less of obliging—authors to abandon their rights in respect of broadcasts to developing countries. These broadcasts, quite obviously, will not be made by the latter countries, but by the industrialized countries and organizations composed of their nationals. It is difficult to think that a court would go so far as to dismiss any applications against the broadcasting organizations—in France this would doubtless be the ORTF in collaboration with other organizations—for the simple reason that it was a question of broadcasts made for the benefit of the third world, however praiseworthy this aim might seem.

The answer which is therefore called for, from a legal viewpoint, would be either a prior agreement between the broadcasting organization and the holder of the copyright with a view to reduced, if not nominal, remuneration of the latter, or the establishment of a fund to meet these expenses. Of course, this fund could be regional or universal, as the need arises.

QUESTION 6

How is direct satellite broadcasting to be correlated with other systems of satellite communications and in particular:
1) What problems does direct satellite broadcasting create in respect of the determination of the original positions of satellites, the allocation of frequencies and how can measures to avoid interference with other transmissions, and measures to limit or prevent spillover, be used in direct satellite broadcasting for the solution of the problems in 1 to 5?
2) What adaptations are necessary to deal with those problems in the ITU Convention and Radio Regulations?

REPLY

The question of correlation of direct or semi-direct broadcasting systems with other systems only arises to the extent where the former exhibit a pronounced specificity in comparison with the others.

The only technical characteristic truly common to direct or semi-direct broadcasting satellites is the fact that they need great broadcasting power.

None the less, for non-technical reasons, direct and semi-direct broadcasting will run the risk of raising problems concerning all the present international regulations on telecommunications by radio waves. Its use as a means of communication in the future to serve developing countries will in the first place lead to an increase in the requests for frequency allocations, without there being any certainty that the frequencies thus allocated will always be used. But it must be emphasized again that the present rule of 'first come, first served', as it is applied to telecommunications satellites, is harmful, and it is most desirable that it be replaced by rules that would in practice serve the general interest.

For these reasons it is greatly to be wished that the assignment of frequencies should be made the object of prior planning, taking into account the needs and technical and economic potentials of the different countries and different regions concerned. It would be a matter of preventing the abuse of frequencies, as well as the reservation of frequencies not being used.

This planning should certainly take into account the probable regional character of the future arrangements for satellite broadcasting. It should also take into consideration the anticipated technical evolution, and reserve the possibility of the emplacement of truly direct-broadcasting satellites whose broadcasts will be capable of being received by individual receivers. In short, it would be advisable to draw up an overall plan for the sharing out of frequencies for a certain number of years, taking into consideration all the factors which might be pertinent: technical evolution, needs of the different countries in telecommunication matters, economic possibilities, necessity of contributing to the development of certain countries, possibility of regional regrouping and linking up of networks, etc.

However, it does not seem that the work of anticipating and of planning will be enough to guarantee the effectiveness of the system; once the plans have been made, they must still be implemented and their application be supervised by a specific body. It is to be foreseen also that in practice problems of coordination will certainly arise, not to mention possible disputes which could spring up between different broadcasters. To solve all these problems it will be necessary to create a special body for the drawing up of plans as well as for supervision and for correlation once these plans have been made. This body should be instituted on a universal level, within the framework of the ITU. It should be permanent and should have a certain independence in relation to the governments of the member states of the ITU, in the same way as has the IFRB at the present time.

It is clear that the creation of a special body of the sort which is envisaged would bring about appreciable changes concerning the very conception of the ITU. It should be emphasized, nevertheless, that these new duties should be taken on by the ITU and not by institutions set up outside it, because of the universal character of the ITU, as well as to preserve a certain continuity which would without doubt be of assistance in facilitating the carrying out of the new duties. Finally, it should be underlined that a universal organization, connected with the UN, as is the case of the ITU, might be particularly fitted, if given the necessary competence, to arbitrate in disputes between its members or groups of members. This last prospect seems particularly important: the creation of broadcasting organizations on a regional basis makes necessary a universal organization which will be able to correlate the activities of the different regional ones, principally on a technical level and perhaps also in other fields. It may only be asked if the present structures of the ITU are adapted to face such a large-scale task, and if, finally, direct satellite broadcasting does not, as a whole, raise the problem of a world-wide organization, endowed with powers and structures adequate to meet the demands of the coming decades: planning, correlation, technical assistance, settlement of disputes. In short, it is the ITU statutes themselves which should be remodelled, to transform the organization by endowing it with more far-reaching and powerful resources.

Some other minor modifications should also be made to the ITU Radio Regulations. Thus, it would be logical to bring certain provisions of the Radio Regulations into line with direct or semi-direct broadcasting An amendment of Article 7, para. 1 (422) of the Radio Regulations, according to which 'the establishment and use of broadcasting stations (sound broadcasting and television broadcasting stations) on board ships, aircraft or any other floating or airborne objects outside national territories is forbidden', seems necessary in order to prevent this text being interpreted as a prohibition on the installation of radio transmitting stations on satellites. Further, it would be advisable to adapt Article 7, s. 1 (2) (423), which imposes limitations on the power of broadcasting stations, to the requirements of broadcasting satellites, which are of necessity very powerful. Finally, one would wish that the provisions of Article 9 (9a) and Resolution 1 A, revised and completed, would be incorporated in Article 9 of the Radio Regulations.

IV REPLY OF THE JAPANESE GROUP

ASSUMPTIONS

1. The Working Group on Direct Satellite Broadcasting of the UN Outer Space Committee has concluded that reception from direct broadcasting satellite with community receivers would be possible by 1975. In fact, the experimental broadcasting project between the United States and India is to be implemented in 1975. On the other hand, reception with home receivers, even augmented, is not considered possible until 1980, while reception with augmented receivers will not be realized until 1985. In the circumstances, it is deemed appropriate here to answer the questions regarding reception with community receivers. It will probably be more practical to consider the question of reception with home receivers after several years of experience with community receivers, and in what follows 'direct' is to be understood to refer to broadcasts to, or reception by, community receivers from satellites.

There will probably be certain differences in technical and physical control imposed on these two modes of reception by the receiving countries. In the case of community receivers, there will be room for making a technical check of broadcasts, but not in the case of home receivers, and this may well raise serious problems. Nevertheless, in theory, a number of problems may be considered in the same light with regard to both modes of reception.

2. In the immediate future, transmission and reception of broadcasts by direct broadcasting satellite are likely to be conducted under bilateral agreements or, if the satellite is to serve a particular region, agreements among countries in that region. In the case of a bilateral agreement, many questions posed in this questionnaire, with the exception of spillover problems, may well be solved within the framework of the agreement. Many problems will probably be solved within the context of multilateral agreements, if the region lies within the same cultural sphere, for example, Europe. Of course, such questions will arise as whether a comprehensive agreement should be concluded, or intergovernmental agreement limited to basic matters with concrete details of implementation left to an agreement between such non-governmental organizations as broadcasting companies. In any case, legal agreements are likely to regulate a considerable portion of the direct-satellite broadcasting service. It goes without saying that the substance of such agreements should be determined by the countries concerned, but the establishment of

some framework for such agreements may be desirable, in order to ensure proper conduct and development of direct broadcasting. Many of the answers submitted herewith are considered useful in indicating standards for such agreements.

Transmission and reception of direct broadcasts under agreements among broadcasters, without the intervention of governments, are not inconceivable, but such agreements should also respect the standards indicated in the answers following. Transmission of broadcasts without any agreement with the countries on the receiving side is also conceivable. Agreements may be necessary, of course, for such broadcasts to be protected and free of interference in the receiving countries, but even without such protection, there is a possibility that broadcasts may be transmitted widely and received in all parts of the world. Global broadcasting by satellite linkage is also possible. Aside from exceptions, such as broadcasts of the Olympic games or United Nations debates, broadcasting with global coverage is likely to be rare, since programmes intended for the developed and developing countries will naturally differ, and problems of language and time differences will continue to exist. Consequently, extensive global broadcasting is likely to be far in the future, if it comes at all. The answers that follow will indicate the standards that should serve as the basis for intergovernmental or interbroadcaster agreements. Problems resulting from the use of direct broadcasting satellites without agreements, including global broadcasting, will also be considered.

3. Aside from broadcasting under bilateral agreements, regional broadcasting will probably form the centre of interest for the time being. In that case, it must be noted that identical answers from the same perspective to the problems posed in the questionnaire would not necessarily be wise because of differences both in the level of development and in the degree of cultural commonalty within the region in question. For instance, no single concept of copyright and neighbouring-right protection with respect to broadcasting by the direct broadcasting satellite would be feasible; the concept required for broadcasting directed to a region composed of developed countries would likely differ from that required for a region consisting of developing countries, particularly in the case of educational broadcasting.

QUESTION 1
What means of promoting and extending freedom of information and communication does direct satellite broadcasting provide, in particular for developing countries?

REPLY

1. *Observance of the Principle of Freedom of Information by Developing Countries*

(1) The principles for regulating direct broadcasting as a whole are not yet established in positive international law, and await implementation in the future in the form of legal rights and obligations. A notable example is the principle of freedom of information and its limits. There seems to be a number of prima facie precedents applicable to this question, but the principle of freedom of information and its limits, referred to in many international documents,[1] can hardly be considered as general international law binding all the nations of the world. It is also highly doubtful whether they are applicable as they now stand to the problems of direct satellite broadcasting. It is not a matter of *lex lata* but of *lex ferenda* to relate the principle of freedom of information to direct satellite broadcasting, and there is room for policy controversies both at home and abroad on judgments of conflicting social values.

(2) With regard to the future application of the principle of freedom of information to direct satellite broadcasting, due attention must first be given to the fact that there will be differences between the developing and developed countries in the significance, content, and social functions of the freedom of information and its limitations. Consequently, it would be particularly important to know the extent to which direct broadcasting will be protected under the national laws of developing countries.

Through the establishment and expansion of direct broadcasting real benefits will accrue to education and culture in the developing countries and friendly relations among nations will be promoted. By these means social development under the principle of freedom of information will be also further enhanced in the developing countries. However, it may be questionable whether, in the process of reaching such a mature stage, the developing countries will, like the developed countries, unconditionally accept the 'freedom to seek, receive and impart information' as important fundamental human rights of individuals. In the developing countries, systems for education and broadcasting to the general public have so far been inadequate. For these countries the introduction of direct satellite

---

[1] For instance, attempts have been made to establish the principle and its limits with the Convention Concerning the Use of Broadcasting in the Cause of Peace (Geneva Convention) of 1936, the Universal Declaration of Human Rights (1948), Convention Concerning International Dissemination of News and Right of Correction (1949), Draft Convention Concerning Freedom of Information (1951), Convention on the International Right of Correction (1962), and International Covenant on Civil and Political Rights adopted by the UN General Assembly on 16 December 1966.

broadcasting as a new information medium will probably arouse concern over its social impact, the intensity of which would be beyond the imagination or prediction of any developed country. It is quite possible that developing countries may invoke various principles under international law, such as sovereign equality, non-intervention in domestic affairs, and national self-determination, and attempts may be made to limit the principle of freedom of information to their own advantage. Also, in not a few cases developing countries may limit or deny the freedom of information under their national laws from the standpoint of public order and morality, national security, and the exigencies of government. Consequently, it would be necessary to protect and guarantee positively the freedom to seek and impart information in the countries concerned to facilitate direct broadcasting.

Aside from the need for mutual cooperation and self-regulation among the broadcasters, guarantees under intergovernmental agreement between the two countries concerned will undoubtedly be required, in order that every aspect of the freedom to 'seek, receive, and impart information' may be realized.

(3) In order that the principle of freedom of information may prevail in the developing countries, the form of organization of direct satellite broadcasting will have to be carefully studied. It might follow one of two possible patterns: a single system on a bilateral or regional level, or a proliferation of competing systems and promotion of free competition among them. Should the former formula be adopted, the broadcasters or the countries having jurisdiction over them would be legally prohibited from establishing, using, and participating in other competing systems (the term 'single' is taken to mean exclusive on the international level). In this case, the basis and conditions for recognizing the de facto 'monopoly' of the direct satellite broadcasting service under the single system would require careful study in connection with the principle of freedom of information. However, in order to realize the principle of freedom of information in the developing countries and to assist their people in the selection of programmes, the second pattern may be considered desirable. In that case, problems will arise of technical and economic adjustment and/or integration of multiple systems.

2. *The Position of the Developing Countries with regard to Establishment and Operation of Direct Satellite Broadcasting*

(1) In so far as the establishment and operation of direct satellite broadcasting can assist the developing countries socially and culturally, as a general principal, the opinions and wishes of any developing country which receives the service must not be ignored. Rather, such countries should be actively invited to take part in the service,

The first step would be to determine the extent to which a developing country's participation and involvement in the establishment and operation of the service could be accepted and guaranteed. In particular, the type of satellite to be launched, the orbit, frequencies and power to be used, and the operating conditions of the directional antennas will greatly affect the technical standards and cost of construction of receiving facilities, the coverage, and the charges for the programmes. Demands may be made for some system that will assure expression of the hopes and opinions of the developing country on the policies underlying programmes and their content.

(2) To this end a study would have to be made to determine whether direct broadcasting should be operated under a bilateral or multilateral agreement, and whether or not it should be managed by some regional organization. In the latter case, the voting procedure would become the focal point of issue. In the absence of any *jus cogens* regulating the establishment and operation of direct broadcasting services among nations, a multilateral agreement or a system based on regional organization will probably be accepted as generally desirable, lest the political and economic power relations existing between developed and developing countries be reflected in the operation of the system.

QUESTION 2

In what ways, institutional or normative, may (a) hostile propaganda, (b) programmes tending to disturb or erode religious, cultural or social life, and (c) advertising, be limited or prevented in direct satellite broadcasting consistent with the principle of freedom both to impart and receive information?

REPLY

1. *Freedom of Information*

(1) Shortwave broadcasting of wide international scope is already in operation to impart useful information, and is valued as a medium for promoting the freedom to receive such information. On the other hand, the extent to which such broadcasting services impair the interests of the receiving countries through transmission of undesirable programmes is believed to be small. This appraisal, indeed, is valid with respect to radio broadcasting. Quite apart from the 1936 convention described later, any claim that institutional or normative limitations should be imposed on existing international shortwave broadcasting for the purposes cited in this Question is not likely to be considered well grounded.

However, it would not be appropriate to consider direct television broadcasting in the same light as radio broadcasting. In the first place, the impact on the audience of television, which appeals to the audio-visual senses, is far greater than radio or other communications media. If, therefore, any harmful programme should be received, the impact on the people receiving it, and ultimately on society as a whole would be relatively great. It is natural that the receiving countries should be concerned with prohibiting or limiting undesirable broadcasts. In the developing countries in particular, where mass media are still inadequate, the impact of television would be strong. It will hardly be possible to hope for sound development of direct broadcasting if such apprehensions of the developing countries are ignored.

In the second place, protection of broadcasting in the receiving country would be necessary in order that television broadcasts should be received by community receivers without impediment. Without such protection, television cannot be a sufficiently effective medium. In the third place, direct satellite broadcasting utilizes a satellite in outer space, which is under international protection, and uses internationally-allocated frequencies. Therefore, it must be considered not as a private, profit-seeking enterprise but as a public service, so to speak, of international character.

In the light of the above considerations, some limitations are needed on direct broadcasting. Complete disregard of its effect on the principle of freedom of information, and unlimited broadcasts of any kind might tend to impair the interests of the receiving country. In that case, no cooperation would be forthcoming for broadcasts, and the sound development of satellite broadcasting might be impeded, both contrary to the public nature of such broadcasting.

(2) On the other hand, direct broadcasting offers a new means of disseminating information; it provides opportunities of enjoying and promoting the freedom 'to seek, receive and impart information' on a global scale. In particular, more importance should be attached to the freedom of the people of the receiving country to receive and enjoy ideas and information rather than on the freedom of the sending country to transmit.

The basic question is how these two seemingly contradictory requirements can be harmonized. Since it will be necessary for both the sending and the receiving countries to compromise to a certain extent for the sake of sound development of satellite broadcasting, let us now consider measures that are most likely to develop the desired harmony.

(3) It is appropriate first to consider the application of existing international law.

Endeavours have been made to establish the principle of the freedom to impart or receive information on an international level. For instance, the UN has adopted the Universal Declaration on Human

Rights (Art. 19) and the International Covenant on Civil and Political Rights (Art. 19), which concern freedom of information. The former, however, did not have, and the latter has not yet, any force, and, except for regional normative conventions with legal binding force, like the European Convention on Human Rights, the principle of the freedom of information cannot at present be considered a rule of general international law. For example, it is questionable whether a deliberate jamming of broadcasting by another country could be adjudged a violation of any rule of international law.

On the other hand, what has been done with respect to restrictions on the freedom of information? The International Convention Concerning the Use of Broadcasting in the Cause of Peace adopted in 1936 deserves attention. This Convention is notable for listing important principles related to programme standards for international broadcasting, but few countries have ratified it; and, though its continuance was confirmed in 1954 by the UN General Assembly resolution 814 (IX), there is little likelihood of any increase in the number of ratifications. The Convention on the International Right of Correction adopted in 1962 provides an additional means of guaranteeing the accuracy of information. This convention recognizes the right of any country concerned to demand correction of false or distorted information. Since this too has been ratified by only a few countries, there is still no legal norm generally prohibiting false statements.[1]

Efforts have been made to limit or prohibit in international law harmful propaganda which might, for example, lead to provocation of war, instigate destructive activities, or defame foreign countries. However, precisely what kind of propaganda can be so described is still not clear, and the relevant rights and obligations arising under international law have not been clarified.

2. *Restriction of the Freedom of Broadcasting*

(1) Effective means of suppressing harmful satellite broadcasting do not yet exist under international law, though efforts are being made to limit or prohibit such broadcasts. The same applies to hostile propaganda, programmes tending to disturb or erode religious, cultural, or social life, and advertising. It would seem appropriate, therefore, to evolve some new means of checking undesirable broadcasts by satellite and controlling their content.

---

[1] GA res. 110 (II) adopted in 1947 disapproves of any propaganda leading to provocation of war, but it remains a mere resolution without any legal binding force. The Outer Space Treaty states that the above resolution is also to be applied to outer space, but the statement is found only in the Preamble. Art. 6 of the Outer Space Treaty might appear to be applicable to the regulation of the contents of satellite broadcasting, but according to our interpretation, satellite broadcasting is not to be considered a space activity as provided in that article.

(2) Even if satellite broadcasting should be conducted under an agreement between the sending and receiving countries, it is desirable, as noted later, for the broadcasters to regulate the content of broadcasts voluntarily by certain ethical standards. It would be possible then in the agreement to cover the various issues raised in the questionnaire, to give legal binding force to provisions made, and to establish penalties for violations. It would be possible to impose reasonable controls on broadcasting programmes according to the mode of broadcasting planned (educational broadcasts, for instance). Moreover, when satellite broadcasting is operated for a specific region under a regional agreement, legally binding regulations may possibly be included in the agreement for controlling programmes. In a region such as Europe, where cultural interests are similar and agreement of the countries concerned can be obtained on standards for programme content, it should not be difficult to use such means to promote rationalization of programmes, either institutionally or normatively.

In devising legally binding regulations for programmes, it should be possible to determine the contents of regulations not only by ethical standards but in light of the wishes and concrete circumstances of the countries concerned, whether the relevant agreement be bilateral or regional. However, since satellite broadcasting will serve the promotion and expansion of freedom of information and is, as stated before, of particular benefit to the people of the receiving country, care should be taken that no unreasonable restrictions on programme content are contained in the agreement. While it is necessary to consider the political system, racial traits, religions, and peculiarities of customs and manners in the receiving country, the principle of freedom of information should also be respected, and without special reasons regulation should be limited solely to what are considered harmful broadcasts under existing international law.

(3) What, in general, would be an appropriate means for prohibiting or limiting such programmes on the lines indicated in the questionnaire? A code of ethics might be established, to be legally binding directly on the broadcasters and so self-executing, or indirectly by government measures of implementation. In any case, prevention of broadcasts proscribed by the code would be considered an obligation under international law.

Such a system would involve institutional means of suppressing broadcasts contrary to the code and would inevitably be attended by a number of difficulties, which could ultimately limit markedly the freedom of direct broadcasting, as well as interfere with beneficial broadcasting in contradiction of the principle of freedom of information. In the first place, international agreement as to what items should be included in the code would be difficult to obtain.

In the second place, though certain terms and conditions may be included, there are extensive possibilities of divergence in their interpretation, and such divergence might well give rise to disputes. For instance, if broadcasting which is deemed intervention in internal affairs is prohibited, the highly ambiguous question of whether a particular broadcast can properly be classified as an intervention in internal affairs would be raised. In the result, the freedom of direct broadcasting might be subject to rigid restrictions.

In the third place, the means of ensuring observance of the code would also pose a problem. Advance control of broadcasts may be highly effective for the code to achieve practical results and protect the receiving country from harmful influences. However, this method would greatly restrict the freedom of broadcasting, and in not a few countries, such as the United States and Japan, it would not be feasible from the standpoint of the constitutional law of the sending country. The mode of subsequent penalty would also be likely to involve considerable restriction of the freedom of broadcasting. Moreover, in countries recognizing the freedom of broadcasting in domestic law, government responsibility for broadcasting will tend to move domestic legislation in the direction of restricting such freedom.

From the foregoing, it appears to be wiser generally to avoid the establishment of any legally binding codes of ethics by international agreement. However, where many of the countries involved, particularly, one might expect, some of the developing countries, refuse to cooperate in the reception of direct satellite broadcasts without such a code, its establishment in a minimal form may be unavoidable. In this case, it would be appropriate to limit the restrictions to broadcasts considered harmful under existing international law.[1] Disputes arising over the enforcement of prohibitive provisions should preferably be settled by some special body similar to the International Committee on Intellectual Cooperation envisaged in the 1936 Geneva Convention (which appears to be the predecessor of the present Unesco), though ultimately the matter might be taken to the International Court of Justice for settlement. Regarding procedure, rules similar to the Rules for the Application of the Convention (1937) should be established (the system of enforcement should be considered separately, however).

(4) Nevertheless, with regard to 'a' and 'b' of the Question, it seems

---

[1] The 1963 Geneva Convention mentioned previously can serve as a useful reference. In it are prohibited broadcasts likely to prejudice good international understanding, broadcasts that tend to incite acts incompatible with internal order or security, and broadcasts whose inaccuracy is or ought to be known. Other types, particularly broadcasts defaming another country, may also be included in the list.

more desirable to work out some concrete international codes of ethics, which broadcasters will observe voluntarily, than to prohibit or limit satellite broadcasting under a code of the type just described. This system is compatible with promotion of freedom of broadcasting, and if properly observed, could be effective in checking undesirable broadcasts. In Japan, harmful broadcasting is successfully checked by regulation based on ethical standards. Moreover, consideration might well be given to directives for satellite broadcasting to contribute to cultural exchange, international goodwill, and the greater welfare of mankind.

However, it is not proper to leave observance of the ethical code entirely to the voluntary regulation of the broadcasters. If voluntary self-regulation in the sending country is the only means of control, it would probably be impossible to dispel the concern of the receiving country. Consequently, it will be necessary to set up some body to supervise observance of the ethical code. Its members might be selected by the broadcasters, keeping it close to a system of voluntary self-regulation; but it might then have difficulty in winning confidence in the impartiality of its judgement. It would be better to establish an independent body, composed of internationally trusted persons of learning and expertise, selected on a world-wide basis, and including members from both the sending and receiving countries. This body would assume responsibility for applying and supervising the code to broadcasting.

If any claim to the effect that broadcasts violating the code have been transmitted should be made by the receiving country, this body would investigate the facts of the case; should a violation be found, the body would publicize the fact widely and issue a warning to the broadcaster against transmitting such programmes. Correction or retraction of the broadcast in question might also be recommended, depending on the case. However, no legal penalty would be possible since the code would not be legally binding.

Further sanctions against the violation in question might well be provided for under the domestic laws of the country in which the broadcaster operates: for instance, the broadcaster may be deemed to have committed an extraterritorial offence under the laws of the receiving country. In such cases, legal measures would be taken in accordance with the laws of the countries concerned. Also, whether any steps would be taken against a community receiver that received the undesirable broadcast, and what sort of control should be imposed on it would fall to be determined by the laws of the country in which the receiver is installed.

(5) Special provision appears to be necessary for advertising, about which the broadcasting systems of different countries are not in perfect agreement. If advertising is freely permitted, its impact on

the broadcasting system of a receiving country may well be considerable. It seems appropriate in general to prohibit advertising in satellite broadcasting, or to authorize it only when its reception is approved by agreement. When advertising is permitted, control of the contents should not be dependent on legal standards; an ethical code such as that described above should be prepared to govern undesirable advertisements.

QUESTION 3

In what ways does the individual require protection in direct satellite broadcasting from (a) defamation or false statements in particular by a right of reply and the duty of rectification, (b) invasion of privacy such as undue publicity of personal life.

REPLY

(1) The answer to Question 2, for the most part, seems applicable to this question as well. The existing legislation of various nations with regard to defamation and invasion of privacy shows broad differences in the choice of factors constituting such offences, in the grounds for exemption, and in methods of rectification. Consequently, it would be very difficult to regulate such matters legally under any international code, which means there is a danger that the matter will be a subject of frequent dispute. Measures, therefore, should be limited to the evolution of ethical standards for respecting the honour and privacy of others, and to clarifying the moral obligation of broadcasters to observe these standards. Observance would be supervised by the body already described. In case of an infringement of human rights, the body would make the fact public, and in addition to issuing a warning, might recommend not only correction or cancellation but also an apology on the part of the sender. The broadcaster would be called upon to conform to such order when he receives it, expeditiously and by appropriate means.

(2) If the aggrieved party should not be satisfied with measures based on ethical standards and demands legal redress, such as damages, steps would be taken under the applicable domestic laws. In other words, the aggrieved party could bring suit in the court having jurisdiction in order to seek redress in accordance with applicable law. To facilitate such relief, it would be desirable to establish some international agreement designating the courts to have jurisdiction in such matters.

(3) In the case of regional direct broadcasting, existence of an almost identical concept of honour and privacy protection is, of course, possible. In such a case, it might not be difficult to conclude an international agreement, capable of legal enforcement, for the pro-

tection of honour and privacy. This approach is likely to be useful in promoting the trend towards broader international recognition of legal remedies regarding defamation and invasion of privacy. Also, if such a regional agreement should be concluded, the establishment of some regional international judicial organ might be envisaged, in addition to the preparation of a substantive law. This arrangement would be desirable since the aggrieved would be provided with an international means for seeking remedy.

QUESTION 4

In what ways, institutional or normative, do authors, performers, and producers of broadcast programmes, and producers of phonograms, need special protection for contributions to direct satellite broadcasts in respect of (a) economic interests in the contribution? (b) the intellectual and artistic integrity of the contribution? Do direct satellite broadcasts themselves need special protection?

REPLY

1. *Economic Interests*

(1) With the coming of direct-satellite broadcasting, the size of the audience for broadcasts of plays, musical performances, and records will be vastly expanded. Of particular importance is the fact that the audience will expand in scope beyond national boundaries and become international. As a result, such broadcast productions and recordings will contribute more than ever before towards international cultural exchange, promotion of education, and dissemination of information. In keeping with such developments, the economic interests of authors, performers, broadcasters, and producers of records will have to be appropriately guaranteed. The protection of these interests will serve to assure more definitely the cultural, educational, and informational contributions to satellite broadcasting. In relation to the developing countries, satellite broadcasting will have special significance and will face special difficulties (this matter will be dealt with in Question 5); however, there will be no basic difference between developing and developed countries on the point that satellite broadcasting will promote and expand exchange and dissemination of cultures, education, and information.

(2) The economic interests of authors and other professionals involved may be secured under the copyright laws of each country and under consolidated and expanded copyright conventions. The same fundamental situation will prevail in the case of satellite broadcasting. It is desirable that all countries subscribe to either the Berne Convention or the Universal Copyright Convention. The Berne Convention, in addition to providing for national treatment

in the protection of copyright, expressly prescribes direct protection for a wide range of copyright covering such matters as reproductions, translations, performances, and broadcasting. The Universal Copyright Convention does not expressly provide for direct copyright protection except in the case of translations, and provides in principle for national treatment in the protection of copyright only. In these circumstances, international protection of copyright may well be inadequate or diluted, given the copyright laws of signatory nations. On this point, it is more desirable to subscribe to the Berne Convention. Lately, however, the revision of the Universal Copyright Convention provides for direct protection of copyright in reproductions, public performances, and broadcasting. This is a welcome trend in light of the problems stated above. As noted later in the response to Question 5, it seems necessary that exceptional provision be made in international conventions for developing nations, but such consideration of their needs should not be instrumental in disturbing the basic system of international protection of copyright.

There are countries in the environs of Japan which remain outside both of the conventions, such as the USSR, the Chinese People's Republic, North and South Korea, Vietnam, and Indonesia. It is desirable that these countries become parties to one or the other of the conventions; but if that is not a likely development, countries closely concerned should endeavour to conclude regional or bilateral copyright protection agreements (or clauses) with such nations. A copyright clause for reciprocal protection in the Japanese Peace Treaty with the Republic of China (Taiwan) is a case in point.

(3) In the case of direct satellite broadcasts, whether to individual receivers or community receivers, copyright clearance should generally be undertaken on the transmitting side. In the case of individual receivers, it should be a matter of course for the transmitting country (or the transmitting station) to clear copyright, and it could follow the existing practice with regard to international broadcasting. In the case of community receivers, the question will arise whether or not the process of transmission from the transmitting station to the community receivers corresponds to 'broadcasting', as specified in legal provisions for national or international copyright. In existing copyright provisions, broadcasting is defined as 'transmission by wireless means for direct reception by the public' (see Art. 2, para. 2, item 8, the Japanese Copyright Law; and Art. 11 bis of the Rome Convention). It is problematical whether reception with community receivers conforms to the 'direct reception by the public' mentioned therein. If reception by community receivers is interpreted as conforming to that definition of broadcasting, copyrights will have to be cleared at the transmitting end. Where

programmes are transmitted by request of a receiving country or station, payment will be made for the programmes in accordance with an agreement, and the copyright fee will be included in the charge. If, however, under existing copyright legislation, this form of transmission should not be deemed to be 'broadcasting', the operators of community receivers would have to clear copyrights; but in fact such clearance of programme transmissions would be impossible without an agreement, and extremely difficult even under an agreement. Therefore, from the standpoint of copyright law, direct-satellite broadcasting will in general have to be considered 'broadcasting', whether reception is by means of home receivers or by community receivers. Copyright clearance on copyrighted materials by the transmitting station would be undertaken pursuant to a contract between the transmitting station and the copyright holders, or associations of copyright holders. Whether reception is to be by individual receivers or community receivers, clearance of copyright, for both foreign works and domestic works, will be based on the copyright convention to which the country of transmission subscribes and on the law it has enacted in compliance with the convention.

Protection of authors, therefore, requires that the transmitting country subscribe to some specific copyright convention. It is most desirable that it be a party at least to the Berne Convention, but still better if it also accept the Universal Copyright Convention. If a choice must be made between the Berne and Universal Copyright Conventions, the first choice should be the Berne Convention; but the difference between the two is diminished by the new provisions in the Universal Copyright Convention, providing direct protection for a certain range of copyrights.

Though the receiving station may not be required to undertake copyright clearance, it is highly desirable that the receiving country as well as the transmitting country be party to some specific copyright convention. Particularly in case of reception without any agreement or contract with the transmitting station, copyrighted materials in the broadcast will be more effectively protected in such receiving countries.

(4) Protection of interests of performers, broadcasters, and producers of phonograms is a problem of so-called neighbouring rights. Neighbouring rights are not so widely recognized as copyright either in the domestic laws of countries or in international conventions, and the economic interests of authors are more adequately protected than those of others. Both domestic laws and international conventions must be strengthened with a view to protecting the neighbouring rights of performers, broadcasters, and record producers in keeping with their expanded opportunities for audience exposure through satellite broadcasting.

2. The Intellectual and Artistic Integrity of Contributions

(1) The intellectual and artistic integrity of contributions by authors, performers, broadcasters, and record producers is best protected by recognition of their 'moral rights'. This fact does not change in the case of satellite broadcasting. The 'moral rights' are expressly recognized in the Berne Copyright Convention (Art. 6, para. 2), though not in the Universal Copyright Convention. 'Moral rights' are also recognized to a certain extent in the Rome Convention on Neighboring Rights (Art. 13(C)ii), but only a few countries subscribe to this Convention. Most countries subscribing to the Berne Convention recognize 'moral rights' in their domestic copyright law, but this is not the case in countries which subscribe only to the Universal Copyright Convention. In countries such as the United States, that do not recognize 'moral rights' in their copyright law, minimum protection of rights corresponding to moral rights is in substance anticipated under general principles of law.

(2) In light of the foregoing factors, it is desirable that all countries recognize 'moral rights' in their domestic laws, and that they subscribe to some convention recognizing these rights. On the other hand, even where 'moral rights' are clearly recognized, it must be admitted that they may be subject to non-discriminatory restrictions in so far as copyright and neighbouring rights are to be utilized for the public benefit in cultural, educational, and information broadcasts.

3. Protection of Broadcasts Themselves

(1) As in the case of terrestrial broadcasting, copyrighted materials may sometimes be used and sometimes not used in satellite broadcasts. If copyrighted materials are used, protection of the broadcast itself can be considered quite apart from the question of copyright protection. But when copyrighted materials are not used, the sole consideration is protection of broadcasting. This again constitutes a problem of recognition of broadcasters' neighbouring rights (already described in 3(A) above), which must therefore be further developed.

(2) A further problem related to the protection of broadcasting will arise, principally in connection with community receivers. As has already been shown, there is a strong possibility that the phases of direct satellite broadcasting up to and including reception by the receiving station will not be deemed 'broadcasting'. For this reason, even when 'broadcasting' is protected under the 'neighbouring right' principle, it may not enjoy protection (leaving aside the question of copyright) against unauthorized use during the phases noted above. As stated earlier in this answer and pointed out again in the answer to Question 6, we are in favour of considering direct satellite broad-

casting—the entire process from up-links to the community receiver —as 'broadcasting' from the standpoint of the ITU Convention and legal provisions concerning copyright.

QUESTION 5

What problems do the issues raised in Question 4 create for developing countries in the reception of direct satellite broadcasts or access to satellites?

REPLY

(1) Satellite broadcasting has special significance for the developing countries from the standpoints of culture, education, and information. On the other hand, the developing countries will face great difficulties in meeting the costs of broadcast programmes and in constructing receiving facilities. With regard to problems arising from Question 4, if copyrights and neighbouring rights are notably restricted or even denied with respect to broadcasts directed to developing countries, these economic difficulties of the developing countries will be either greatly diminished or dissolved. This will result in free and adequate enjoyment by the developing countries of cultural, educational, and information services that can be useful in enhancing their national development.

(2) However, this should not be considered as implying any tendency towards the general lowering of international and domestic standards with regard to protection of copyright and neighbouring rights. It is true that the United States in the nineteenth century hesitated to recognize international protection of copyright based on its position as a developing country, and that Japan in the Meiji era attempted to restrict foreign copyrights with regard to translations, in part from the standpoint of a culture-importing nation. However, these positions had no controlling influence on the overall development of copyright-protection systems and came under considerable attack both internationally and domestically.[1] Satellite broadcasting is not a problem for the developing countries only. It will be a different but no less important problem for the developed countries as well. Moreover, in the future, developing countries will no doubt gradually assume the position of transmitting countries.

(3) Even though unilateral favours or reservations may be offered for the benefit of the developing countries with respect to copyright, they must be sufficiently reasonable to be acceptable to the developed countries and to the authors and others interested in those countries. This is readily understandable in light of the failure of the Stockholm Act of the Berne Convention. Some convention designed to make

[1] M. M. Kampelmann, 'The United States and international copyright', *American Journal of International Law*, Apr. 1947.

possible and to facilitate participation of both the developing and developed countries should be worked out. Following the failure of the Stockholm Act, the International Copyright Committee and the Permanent Committee of the Berne Union have been jointly studying the problem and preparing revisions of both the Berne Convention and the Universal Copyright Convention. These revisions envisage detailed limitations on the reproduction and translation of copyrighted materials, and a transitional recognition of their unilateral reservation to the developing countries, no reciprocal claims from the developed countries being recognized. Also envisaged is the recognition of transfer of participation from the Berne to the Universal Convention by temporary suspension of the Berne Protection Provisions (declaration attached to Article 17 of the Universal Convention) for the sake of the developing countries (revision of the attached declaration 'a' to Art. 17); or the removal of impediments to the application of the Universal Copyright Convention to member countries, which are at the same time parties to the Berne Convention (revision of the attached declaration 'b' to Art. 17, or of the corresponding part of the Berne Convention). In principle, such developments should be supported.

(4) Utilization of the above systems will enable the developing countries to use foreign works in their domestic broadcasting, including satellite broadcasts, at comparatively low cost. In the case of developed countries' satellite broadcasts to developing nations, clearances of copyright should be carried out under such a treaty system on the basis of special reservations for the developing countries, if the principle of clearance by the transmitting countries is to be maintained.

(5) Careful consideration should also be given to the benefit of developing countries not only with regard to reproduction and translation rights but also in connection with copyrights and neighbouring rights related to performances and broadcasting.

QUESTION 6

How is direct satellite broadcasting to be correlated with other systems of satellite communications and in particular:
(i) What problems does direct satellite broadcasting create in respect of the determination of orbital positions of satellites and the allocations of frequencies, and how can measures to avoid interference with other transmissions and measures to limit or prevent spill-over be used in direct satellite broadcasting for the solution of the problems considered in questions 1 to 5?
(ii) What adaptations are necessary to deal with these problems in the ITU Convention and Radio Regulations?

REPLY

1. *Features of Direct-Satellite Broadcasting Service*

(1) Satellite communications services in general imply point-to-point communication, and the communications satellite largely plays the role of a transmission satellite.[1]

Telecommunication laws in general have been framed around transmission of communications. In other words, the communications common carrier has made it his primary responsibility to undertake terminal-to-terminal control of telecommunications, so that continued quality and reliability will be ensured in point-to-point communications. The situation outlined above does not change with regard to general space communications.

(2) On the other hand, direct satellite broadcasting uses space stations for transmitting or retransmitting signals which are intended for direct reception by the general public; the space station is thus the focal point for high-power emission.

In this service, therefore, it is necessary to ensure the integrity of programme content during the whole process of broadcasting from the original transmission to the direct reception by the general public. In this one can see the international public-service character of the direct-broadcast satellite which distinguishes it from other types of communications satellite.

2. *The Use and Allocation of Geostationary Orbits and Radio Frequencies*

(1) From the standpoint of the actual supply and demand, it is still premature to designate the geostationary orbit a limited natural resource; it would therefore be risky to establish a formula for allocating broadcasting rights and obligations among countries for the use of geostationary orbits, based on such an assumption.

(2) Continuous use and occupation of a geostationary orbit by a broadcasting satellite does not come within the purview of the national appropriation outlawed by Article 2 of the Outer Space Treaty. Actual occupation of a geostationary orbit by a satellite does not inevitably exclude the use of the orbit by another satellite, based on adjustments of frequency and antenna direction; nor is such exclusion necessary, at least under presently envisaged conditions of demand. That is to say, occupation does not of itself involve a claim

---

[1] For instance, the WARC–ST convened by ITU in 1971 defined the fixed-satellite service as radiocommunication service between earth stations at specified fixed points when one or more satellites are used; in some cases this service includes satellite to satellite links, which may also be effected in the inter-satellite service, and for connection between one or more earth stations at specified fixed points and satellites used for a service other than the fixed-satellite service (for example, the mobile-satellite service, broadcasting-satellite service, etc.).

of right to exclusive use based on territorial control and jurisdiction in outer space.

(3) Consequently, the allocation of rights in the use of a geostationary orbit should be made under a regime of coordination based on international consultations. The states themselves enjoy freedom of the use of outer space on equal footing, in accordance with the provisions of the Outer Space Treaty (Arts. 9 & 1). In so far as this procedure is followed, allocation of the use of geostationary orbits will be analogous to the system of frequency allocation in both technical and legal contexts. In addition, it would seem most appropriate to decide upon the allocation of use of geostationary orbits within the framework of the ITU, utilizing the coordinated frequency plan.

(4) In connection with frequency allocation, the ITU will probably be faced with a need to undertake reallocation of frequency bands, in line with the development of new radio-communications services. Fundamental reconsideration may have to be given to the question of whether or not the ITU should maintain the existing regime by which previously allocated and registered frequency bands are matters of de facto vested right, and a virtual right of priority is given to frequencies registered with IFRB, under the first-come, first-served rule. The existing machinery of frequency allocation and use on a shared basis by service and region might have to be amended with regard to direct-satellite broadcasting.

(5) The members of the ITU are obliged to prevent harmful interference with the radio services of other countries caused by private operating agencies under their jurisdiction, under the principle of state responsibility under international law. So far, it has sufficed in practice for the members to be diligent in taking such steps as they deem necessary under their national law to ensure that private enterprises observe the provisions of the Convention and Regulations related to the prevention of harmful interference (Art. 22, para. 2, ITU Convention of 1965). However, the system of state responsibility under the present Convention should be amended and reinforced with regard to prevention of interference with frequencies to be used for direct satellite broadcasting. The state should preferably bear direct responsibility for prevention of harmful interference caused by private enterprises in view of both the nature of direct satellite broadcasting as an international public service and the principle of state responsibility prescribed in Article 6 of the Outer Space Treaty. Article 7 of the Treaty deals with harmful interference as a mode of damage. However, this would pose the problem of whether such harmful interference should itself correspond to the type of damage requiring compensation under an Outer Space Liability Agreement. This question is now under consideration by the UN Legal Sub-committee for Peaceful Uses of Outer Space.

3. Amendment of the ITU Convention and Regulations

(1) The relevant definition referred to in the present Radio Regulations should clearly reflect the public interest in direct-satellite broadcasting as a whole, and distinguish it from the communication-satellite service now in operation. It is still too early to say whether the new definition of the 'Broadcasting-Satellite Service' in Article 1, 84 AP of the Radio Regulations provides substantive backing for frequency allocation or for a definition of the space station for operating the service.

(2) Study may be needed to determine whether efficient operation of the direct-broadcasting service will be achieved by the allocation of frequency bands for primary uses on the traditional basis of shared use with other space services, or by the establishment of exclusive frequencies.

(3) The existing provisions prohibiting establishment and operation of extraterritorial broadcasting stations (Art. 7, 422, Radio Regulations) should be clarified with regard to the direct-satellite broadcasting service.

V REPLY OF THE UNITED KINGDOM GROUP

Direct-satellite broadcasting is here understood generally as meaning broadcasting for public reception by *community* receivers, and as excluding broadcasts on military or other government service.

The answers to the Questionnaire which follow rest on the assumption that, according to the generally accepted timetable, direct broadcasts to community receivers, and even to augmented home receivers, will be practicable within a very few years, but that reception by existing types of receivers unaugmented is unlikely for a decade or more.

It is also assumed that direct-satellite broadcasting will be limited by time-zone, as well as cultural and language, differences so that global programmes, designed for simultaneous reception in all countries, will be so rare that they can for present purposes be ignored. It is assumed then that direct-satellite broadcasting, as envisaged in the Questionnaire, will be designed largely for regional and national reception.

QUESTION 1

What means of promoting and extending freedom of information and communication does direct satellite broadcasting provide, in particular, for developing countries?

REPLY

Direct-satellite broadcasting could be a means of extending freedom of information, as it is understood in a number of international instruments. However, the scale of its use for spreading information will be determined by many factors, political, social, and economic. In the industrialized countries, and particularly those such as Japan and some European countries, which are densely populated and relatively small in terms of satellite 'visibility', direct satellite broadcasting may be less used as an extended information service than in the developing countries. However, the EBU proposal for narrow-beam direct broadcasting would permit transmission to individual countries, and its use would depend upon national decisions on the control of such a service and the content of its programmes. In the existing communications-satellite system, telecommunications, including data-processing, have been the main use, with broadcasting absorbing less than 5 per cent of the capacity. An intermediate state of development between the present system and direct broadcasting to individual receivers is direct transmission from satellite to community receivers. The final stage of satellite broadcasting may well be the combination, as circumstances require and national authorities permit, of all three systems, all interconnected and providing access to any transmitted material. The community receiver, from which individual receivers could be fed by cable, and the individual receiver itself would be capable of reception on many channels, some national and some international. These could receive signals direct from a national or regional satellite; or from an earth station linked with an intercontinental-satellite system; as well as from terrestrial-broadcasting systems as now, where these remained the national economic choice. The same satellites which provide these broadcasting services could, of course, also be used to send and receive telecommunications of all kinds: the proportion of use between broadcasting and telecommunications would depend on the governmental and financial conditions under which the systems were operated.

Education is by far the most important service which broadcasting can provide in developing countries, and cost effectiveness is a crucial determinant for both the governments concerned and international agencies initiating or supporting educational broadcast programmes. Direct satellite broadcasting could be used to transmit educational TV programmes to remote areas of large countries such as India, Brazil, or the Zaïre Republic (Congo/Kinshasa) and to areas where at present little or no education of any kind is available. However, there are a number of difficulties to meet.

First, electricity must be available in the reception areas and community receivers would generally have to be provided. Generators for local electricity supplies and receivers could be supplied by government grant, supported perhaps by international aid. Secondly, there are familiar doubts about the efficacy of educational TV programmes in schools and institutes, unless there are teachers or group leaders trained and available to elaborate a programme with explanation, questions and discussion. It is suggested that video-cassettes, linked to community receivers by cable TV (CATV), would be more economic and effective than direct satellite broadcasting.

However, these difficulties are not insurmountable, nor is CATV obviously competitive. Even in the United Kingdom, its future usefulness is doubted because of the high cost of cabling in a country where the percentage of homes equipped with telephones is remarkably low. In developing countries CATV is likely, unless very heavy expenditure is incurred, to be capable of very limited use in densely populated localities. Further, the Satellite Educational Television

Experiment in India has been undertaken after a thorough analysis of the needs and problems, and it may well show how the major difficulties of such a service can be overcome.

It is possible to organize collaboration on a world scale for the distribution of news and news analysis by satellite. This is shown by the experience and achievement of EBU in developing its news transmissions. The method used is the selection by news editors in a daily 'closed circuit' conference of news items on film or tape, which are distributed by land circuit or satellite in fixed transmission periods or as 'flash' transmissions. News exchanges have already been conducted within Europe, subsequently both incoming and outgoing with the International Radio and Television Organization and by satellite from North America. Experimental exchanges are now in progress with Latin America, via the earth station in Spain. There could thus be organized extension by direct satellite broadcasting of news and news analysis to and from developing countries, as well as effective news coverage over large national territories.

Reference has already been made to international declarations and conventions on freedom of information. While they serve to establish standards and provide directive principles,[1] they do not by themselves resolve the problems to be faced in the spread of information, whether of news or educational. If we consider an instrument actually in force—the European Convention on Human Rights—we see that the freedom of information which it guarantees is in fact much qualified and hedged with the kind of restrictions, which are in common use and are perhaps inevitable.

1. Everyone has the right to freedom of expression. This right shall include freedom to hold opinions and to receive and impart information and ideas without interference by public authority and regardless of frontiers. This Article shall not prevent States from requiring the licensing of broadcasting, television or cinema enterprises.

2. The exercise of these freedoms, since it carries with it duties and responsibilities, may be subject to such formalities, conditions, restrictions or penalties as are prescribed by law and are necessary in a democratic society, in the interests of national security, territorial integrity or public safety, for the prevention of disorder or crime, for the protection of health or morals, for the protection of the reputation or rights of others, for preventing the disclosure of information received in confidence, or for maintaining the authority and impartiality of the judiciary.

Direct-satellite broadcasting will serve to extend on a large scale not only the sources of news but the audiences to which transmission is addressed. The problems then of maintaining a proper balance be-

[1] e.g. Universal Declaration on Human Rights (1948), Art. 19, Covenant on Civil and Political Rights (1966), Art. 19, European Convention on Human Rights (1950), Art. 10, Interamerican Convention on Human Rights (1969), UN General Assembly res. 2448 (XXIII).

tween the principle of freedom of information and desired social objectives can only be intensified.

QUESTION 2

In what ways, institutional or normative, may
(a) hostile propaganda;
(b) programmes tending to disturb or erode religious, cultural or social life;
(c) advertising

be limited or prevented in direct satellite broadcasting consistent with the principle of freedom both to impart and receive information?

REPLY

The problems which could arise under these heads would be solved if direct satellite-broadcasting systems were under the control of the government of the receiving country and were integrated by it into the national system of licensing and control of all forms of broadcasting. The government might license an existing broadcasting organisation to operate a direct-broadcast service or create a new body or bodies to do so. The extent to which freedom in news broadcasting would co-exist with governmental control of 'harmful' content would then depend on the political position taken by the national government. In the UK, the broadcasting organizations take responsibility for their programme content, within the conditions of a Charter and Licence in the case of the BBC and of the Television Act in the case of Independent Television. Their news services are not subject to government control or supervision.

Thus responsibility for public broadcasting in the United Kingdom is vested in the BBC, established by Royal Charter of 1926, since renewed,[1] which is a legislative instrument. A public TV service, provided also by the BBC, was introduced in 1936; but a second TV broadcasting service was introduced in 1954 under an Independent Television Authority (ITA), established by the Television Act, 1954.[2]

The BBC operates under a Licence and Agreement from the Postmaster General under Wireless Telegraphy Act 1949, and dated 7 July 1969. The office of Postmaster General was abolished as from 1 October 1969, and his functions in relation to broadcasting were assumed by the Minister of Posts and Telecommunications. The current Licence and Agreement expires in July 1976. Under s.13 (4) the Minister has a reserved power to require the Corporation to refrain at any specified time or at all times from broadcasting any

[1] In June 1964 until July 1976 (Cmnd 2385).
[2] Extended until July 1976 by Television Act 1963.
[3] Post Office Act 1969.

matter or matters of any class. This power has been occasionally exercised with respect to certain classes of political broadcasts, in particular, to secure the independence of parliamentary debate. Under s.19(1) the Minister may, in a state of emergency, take control of stations of the Corporation and prevent the Corporation using them, if he believes that the public interest requires it. The BBC is precluded from receiving any money or valuable consideration for broadcasting or refraining from broadcasting anything (Licence, Clause 12).

Independent television broadcasting is conducted by the ITA under the Television Act 1964, which consolidates the earlier Acts of 1954 and 1963. The ITA broadcasts programmes supplied to it by producer companies under contract. Under Television Act 1964 s.4, the ITA is required to draw up, and revise from time to time, a code giving guidance on broadcasting practice and establishing standards, and special provisions govern advertisements, to be considered below. Under s.18 (3) the Ministry of Posts and Telecommunications may at any time require the Authority to refrain from broadcasting any specified matter or classes of matter.

Access by the UK broadcasters to the existing Intelsat communications-satellite system is obtained through Goonhilly Down, the UK earth station owned and operated by the Post Office Corporation; or through the four other West European earth stations in France, Germany, Italy, and Spain, on the tariffs and conditions offered by the operating authorities.

If it were decided to use direct-satellite broadcasting to provide one or more additional sources to the three existing channels, it would probably be brought within the system of responsibilities laid on the BBC and the ITA. The existence now of two news services, offered by the BBC and by the ITN, and drawing on separate editorial and news gathering talent, has affected the amount of information available to the public and has given it a choice of reception times. This could also be the effect of one or more addtional channels available to the viewer direct from a satellite service, if the conditions laid down for the service included news broadcasting and other informational programmes.

In general, then, the use and impact of direct-satellite broadcasting in terms of the Question posed will depend in large part on the system of control.

In perhaps most developing countries the broadcasting operators would be primarily government departments or agencies. In this case the control of news and information could be directly with government and there are, of course, many countries where this is the case now under existing systems. Alternatively the government could assign the provision of a news and information service, with or without other categories of programmes to be included in its transmissions, to a public or private broadcasting organization. The broadcasting operators could be expected to seek programme material, for the programme purposes required, not only from their own production resources but by purchase of film and video tape, which would be used in making up the programme schedule to be transmitted by their direct broadcasting system. Live incoming material might also be available to them via communication satellite for retransmission on their broadcasting system, as, for example, in the case of a live or deferred broadcast of a major international event. Both the freedom to provide information, and controls upon its content, would depend upon the kind of service which the government is seeking to provide, and the assumption made in this paragraph is that the government will be in a position to control any national direct broadcasts transmitted within its country.

Many programmes of all types are transmitted in both unilateral and multilateral arrangements under the auspices of such international organizations as the EBU and OIRT. But the decision to transmit a programme originating in another country lies solely in the hands of each national broadcasting organization.

The EBU may negotiate the television rights to a major sporting event like a World Championship (to be exact, when such an event occurs within Europe, the negotiation is carried out by the EBU member in the country concerned, on behalf of the EBU) and the contract may contain restrictions on, say, advertising placards in the view of the cameras. When a breach occurs, the response and the redress is basically a national matter even though there is a clear EBU policy on this point. There may be a collective boycott by EBU, but there may not be. This usually depends on whether the 'host' broadcasting organization feels strongly about the matter.

EBU members collectively agree on common recommendations, but each individual member decides whether or not to follow them in a given case. But either way the response is straightforward, for a member will not broadcast programmes if in its view they offend against points (a), (b), or (c) in Question 2.

National control may, however, tend to be diluted in developing countries, if they have access to a common satellite under regional agreement. A satellite could be positioned and allocated a frequency to transmit a direct broadcast service which extends beyond national borders to cover a number of countries on a 'regional' basis. The establishment of such a regional system should be the subject of an agreement between the governments of the countries concerned, if a common policy is to be adopted on the extent to which news and informational broadcasting is to be independent of state control and to which any restrictions are to be formulated with reference to 'harmful' content. It must be recognized that such agreements and such definitions are far from easy to achieve and it is possible that national control in developing countries could become diluted within a direct-broadcast service operated under a regional agreement.

A great deal of public and professional discussion has centred on the very different situations which would arise if direct broadcasts were directed to a country by a foreign service not within any regional agreement and if viewers in that country were enabled to receive the programmes; or if a commercially-based 'carrier' system were created which could be booked by any user for a transmission to any audience area. In these situations only an international agreement at the inception of such services to exclude political broadcasting could prevent their use for 'propaganda'; or only such an agreement could attempt to lay down principles which would exclude 'damaging' content.

The problem of the prevention of broadcasts from places outside the territory of the United Kingdom, which are designed for public reception there and constitute an interference with the broadcasting services or a nuisance, is in part resolved by the Marine Broadcasting (Offences) Act 1967, giving effect to the provisions of the European agreement on the subject.[1] Two points in this Act may be noticed here: first, the Act would not in its present terms cover satellite broadcasts. In any case the methods of prevention of broadcasts envisaged in the Act could hardly be applied effectively to satellite broadcasts.

To offer now answers to the particular Question:—

As to (a), decisions on broadcasting of political propaganda would have to be taken within the system of control. So in the United Kingdom, the extent of political broadcasting is limited by agreements between the broadcasters themselves and the main political parties, and in part by law.

The common-law offence in the United Kingdom of uttering seditious words has, it appears, two necessary elements: that the language used is in fact likely to provoke or induce action in 'a matter of State', and that such action is violent or invokes public disorder.[2] In assessing these elements, the court may take into

[1] European Agreement for the Prevention of Broadcasts transmitted from Stations outside National Territories 1965 (European Treaty Series No. 53: 4 *International legal materials* (1965) 115). The Representation of the People Acts 1949–69 cover broadcasts from a station outside the UK of material intended to influence an election taking place in the UK.

[2] In his charge to the jury in *Rex* v. *Aldred (1909)* 22 Cox I: Central Criminal Court, Coleridge J. said: '... whoever by language, either written or spoken, incites or encourages others to use physical force or violence in some public matter connected with the State, is guilty of publishing a seditious libel'.

account the state of public feeling, and the place and mode of publication. In short, the law concerning sedition corresponds closely with the restrictions permitted for the prevention of disorder and crime, and in the interests of public safety, in Article 10 (2) of the European Convention on Human Rights. Broadcasts, which could be politically subversive in foreign countries, would be (1) unlawful if they contained incitement to the assassination of the head of state or the murder of political leaders; (2) controllable in any case by the Minister of Posts and Telecommunications by exercise of his power of prohibition of broadcasts by the BBC or ITA, already described. Prosecution for seditious publications in any form have been very rare in the United Kingdom and none is recorded in respect of any broadcast.

As to (b), under the common law, blasphemous publications in any form will be unlawful only in so far as they tend to cause a breach of the peace or civil strife, a rule conforming with the provisions of Article 9(2) of the European Convention, permitting restrictions prescribed by law in the interests of public safety and for the protection of public order.

Undue interference by broadcasts with cultural and social life in the United Kingdom could be controlled in respect of the BBC by the powers of prohibition of the Minister already described, and in respect of the ITA by the Television Act 1964. Under the Television Act 1964 s.3 (1):

It shall be the duty of the [Independent Television] Authority to satisfy themselves that, so far as possible, the programmes broadcast by the Authority comply with the following requirements, that is to say—
(a) that nothing is included in the programmes which offends against good taste or decency or is likely to encourage or incite to crime or to lead to disorder or to be offensive to public feeling;
(b) that a sufficient amount of time in the programmes is given to news and news features and that all news given in the programmes (in whatever form) is presented with due accuracy and impartiality;
(c) that proper proportions of the recorded and other matter included in the programmes are of British origin and of British performance;
(d) that the programmes broadcast from any station or stations contain a suitable proportion of matter calculated to appeal specially to the tastes and outlook of persons served by the station or stations and, where another language as well as English is in common use among those so served, a suitable proportion of matter in that language; and
(e) that due impartiality is preserved on the part of the persons providing the programmes as respects matters of political or industrial controversy or relating to current public policy.

As to (c), it should be noted that no conditions are imposed on the use of satellites for advertising in Intelsat rules, though there exist codified rules on advertising generally in a number of countries.

Broadcast advertising is limited in the United Kingdom to programmes of the ITA. Television Act 1964 s.7 (3) provides that: 'It shall be the duty of the Authority to secure that the provisions of Schedule 2 to this Act are complied with in relation to advertisements included in the programmes broadcast by the Authority.' The Schedule contains rules as to advertisements and the Minister of Posts and Telecommunications may, by statutory instrument, make regulations amending, repealing or adding to the provisions of the Schedule.

In general the great extension of transmission of programmes which direct-satellite broadcasting brings with it must multiply the situations in which the problems envisaged in (a) and (b) of this Question arise, as has already been observed on Question 1.

The UN Civil and Political Rights Covenant, embodying a principle set out in an early General Assembly resolution, requires that 'Any propaganda for war shall be prohibited by law' (Art. 20(1)). If and when the Covenant comes into force, countries ratifying it must legislate to prohibit such propaganda. But it would not be easy to determine what has to be prohibited. Is reasoned advocacy of a defence programme propaganda for war? Or if, to be propaganda, advocacy must be irrational, biased or false, is it to be prohibited if it incites to war in any circumstances, or only if it advocates war as an instrument of policy, or only if it incites to war against a particular country? The Covenant also permits restrictions on freedom of expression and information, which are provided by law and are necessary 'for the protection of national security or of public order (ordre public) or of public health or morals' (Art. 19(3)b).[1] The breadth of these exceptions is such as not only to make common standards hard to realize, but to permit restrictions in practice of the freedom of expression and information which can in effect nullify it.

To subject direct-satellite broadcasts to international norms of this kind in order to solve the problems posed under (a) and (b) of this Question, would then be quite unrealistic. Propositions for an international 'code of conduct' have indeed been discussed by the Working Group of the UN Space Committee; however, the relevant resolution adopted by the UN General Assembly emphasizes cooperation on both the governmental and non-governmental level as a realistic way of solving these problems. At the present time a 'Declaration of Guiding Principles' is in preparation within Unesco, for submission to the General Conference of member states in autumn 1972. An advisory meeting of experts, including broadcasters from most areas, which was called by Unesco in October 1971, found much to condemn in the first test, which was peremptory and re-

---

[1] Almost identical language is used in the analogous provisions of the American States Convention on Human Rights (1969) Art. 13(2).

strictive to the point of censorship. A widely held view was that broadcasters themselves must hold the responsibility for the content of direct broadcast programmes. The UK Group takes the view that in general, the problems posed in this Question are likely to be best solved neither by governmental controls based upon an international convention or code, nor by the bare self-restraint of individual broadcasters. But principles governing programme content might be adopted by broadcasters on a regional basis. Mutual interest has already led to international collaboration between them, and new regional associations of broadcasters are evolving. The EBU has been established for many years; it has thirty-three members in the European Broadcasting Zone and sixty-three associate members and continually extends its contacts outside Europe. The USSR and neighbouring countries have established OIRT as a joint organization, which has technical links with EBU. The Asian Broadcasting Union is growing in influence, as is also the Union of African National Radio and Television, and an association of Latin American broadcasters has emerged. For direct-satellite broadcasting, these associations could serve as forums of discussion on programme content, and make collaboration possible under working rules both on programme production and on reception of programmes across frontiers. At an international meeting of broadcasting unions in March 1972, a declaration concerning the use of communication satellites was in fact adopted (see Appendix 3). Ultimately, in the view of EBU, the problems must be solved by the conditions of the 'franchise' of the system, including governmental control over reception facilities. Ownership of the direct-broadcasting service might be national or vested in a regional association.

QUESTION 3
In what ways does the individual require protection in direct satellite broadcasting from:
(a) defamation or false statements, in particular by a right of reply and the duty of rectification;
(b) invasion of privacy such as undue publicity of personal life?

REPLY
Much of what has been said in answer to Question 2 is applicable here. In the United Kingdom, broadcasts are subject, like other forms of publication, to the law regarding defamation. Children are protected by restrictions on persons under the age of sixteen taking part in broadcast performances, and on broadcast reports of court proceedings involving them,[1] provisions conforming with the require-

[1] Children and Young Persons Act (1963) ss. 37(1) & 57(4).

ment of Articles 8 and 6 respectively of the European Convention. Direct-satellite broadcasting does not appear to present any new problems here. However, its extendedness may create the need, and certainly the opportunity, for some rethinking on (a) of this Question. In most countries defamation is ground for a civil action for damages, or even in some circumstances for criminal proceedings. Nevertheless, the law relating to defamation can, as in the United Kingdom, be more restrictive than is necessary of freedom of the press and of information.

Further, the law of defamation has in general grown up in the context of relatively small communities, certainly not larger than the nation. It may be asked then, first, whether its general principles of law do not need some modification in the greatly extended 'community' to be served by direct-satellite broadcasting, and in particular whether in many situations a right of rectification or reply is not an adequate substitute for award of damages. In any case, it would be desirable to secure agreement on at least a regional basis on the determination of jurisdiction and the competent forum in actions for defamation: for example, it would be a convenient solution for any court to be competent to award damages or order publication of a reply or rectification, provided that either the normal conditions of jurisdiction were met, or that it was a court of a country which was both the domicile of the plaintiff and a party to the regional satellite broadcasting agreement.

QUESTION 4

In what ways, institutional or normative, do authors, performers, and producers of broadcast programmes, and producers of phonograms, need special protection for contributions to direct satellite broadcasts in respect of
(a) economic interests in the contribution;
(b) the intellectual and artistic integrity of the contribution?
Further, do satellite broadcasts themselves need special protection?

REPLY

Where programme material originating in one country is used for transmission to viewers in another country, supplementary payments may be accorded. Scales of additional payments are laid down for sales and use in the various national markets throughout the world. Precedents exist under EBU agreements, based on negotiations with performers' unions and with copyright owners or societies administering copyrights. These supplementary costs tend to discourage the use by TV broadcasters of programmes from other countries, given also the obstacles presented by different programme standards and audience acceptances. It is clear that pressures by artists' unions and copyright owners on broadcasters of programmes will not diminish with direct-satellite broadcasting.

Direct-satellite broadcasting as such does not appear to present any case for special copyright protection. Authors, performers, and producers of programmes, used in such broadcasting, will have the same protection in law as already exists for use of their work in TV or radio broadcasting. Performances will be licensed for use when the programme is made. However, there are practical limitations. At present redress for unauthorized use of programmes, in whole or in part, or for non-payment of agreed dues, is obtainable through the courts of the seat or domicile of the defendant broadcaster, but, as has been suggested above, these means of redress would need extension to give adequate protection on the scale of direct-satellite broadcasting. Further, the intellectual or artistic integrity of a contribution to a broadcast programme can only be safeguarded by the contributor before the event: for example, where 'no editing' is made a condition of the contract, between contributor and broadcaster, the intellectual or artistic integrity of the contribution is unlikely to be abused. But protests after the event rarely have much effect, and this would obviously be the case with direct-satellite broadcasting.

Its development then could be an opportunity for extension, both geographical and substantive, of the relevant international conventions. In particular, a redefinition of broadcasting would be useful as well as elaboration of the notion of neighbouring rights.

The European Agreement on the Protection of Television Broadcasts (1960) and its Protocol (1965) are territorially limited to countries which are members of the Council of Europe, not including their overseas territories. Article 1 recognizes:

In the territory of all Parties to this Agreement, the right to *authorize or prohibit*:
(a) the re-broadcasting of such broadcasts;
(b) the diffusion of such broadcasts to the public by wire;
(c) the communication of such broadcasts to the public by means of any instrument for the transmission of signs, sounds or images;
(d) any fixation of such broadcasts or still photographs thereof, and any reproduction of such a fixation; and
(e) re-broadcasting, wire diffusion or public performance with the aid of the fixations or reproductions referred to in sub-paragraph (d) of this paragraph, except where the organization in which the right rests, has authorized the sale of the said fixations or reproductions to the public.

Article 3 permits reservations, of which the United Kingdom for example has availed itself, so that protection may be withheld in certain cases. These Agreements have in any case had only limited acceptance in Europe.

The firm assertion of governmental control in these Agreements, and the limited area which they in effect cover, suggest that even regional agreement on the protection of broadcasts is unlikely to be easy when direct-satellite broadcasting is introduced. Article 1 lays down the broad principles which must apply in the protection of broadcasts from unauthorized use, but it is not clear how they could be internationally enforced in practice.

QUESTION 5

What problems do the issues raised in Question 4 create for developing countries in the reception of direct-satellite broadcasts or access to satellites?

REPLY

The problems of programme protection in direct-satellite broadcasting to developing countries need special attention.

As regards economic interests in contributions, two questions arise, as to the form and scale of dues to be paid to the copyright owner, and the availability of funds from which to make these payments,

It is possible that educational programme contributions would be contracted for on an 'all-rights' basis in return for a single cash payment, rather than on a royalty basis common in publishing. Further, it is likely that, at least in the early stages of direct-satellite broadcasting, dues payable to authors, performers, and producers of programmes would form a relatively small part of the total cost of bringing programmes to community receivers in schools or elsewhere. It may be then that the problem is smaller than it first appears.

However, the introduction at Stockholm in 1967 of a protocol to the Berne Convention, which permits developing countries to grant less extensive copyright protection than the industrialized countries, met with strong opposition; and, even if the 1971 revisions of the Berne Convention and the Universal Copyright Convention make certain concessions to the developing countries, the problem cannot be dismissed. In particular, it is arguable that current affairs programmes and 'live' news, and ideally many educational programmes, should in developing countries be exempted from copyright protection, or at least payment of dues, where this supplementary cost might diminish the availability of the programmes.

It would come well within the province of Unesco to organize an international fund to meet such costs. Alternatively it might, in co-operation with the World Bank, devise schemes by which access to satellite broadcasting by developing countries for educational programmes would be made possible by long-term loans.

QUESTION 6

How is direct-satellite broadcasting to be correlated with other systems of satellite communications and in particular
(i) what problems does direct-satellite broadcasting create in respect of the determination of orbital positions of satellites, the allocations of frequencies, and how can such measures to avoid interference with other transmissions, and measures to limit or prevent spill-over, be used in direct-satellite broadcasting for the solution of the problems considered in Questions 1 to 6;
(ii) what adaptations are necessary to deal with these problems in the ITU Convention and Radio Regulations?

REPLY

The Outer Space Treaty, Article 2, recognizes the principle that the space in which satellites orbit the Earth is common and must be shared, and that undue mutual interference between satellites must be avoided. The radio spectrum has also come to be seen as a natural resource, which is common to all and must be conserved for optimum use; and therefore frequencies must be allocated on a basis of wise management, so that band-widths are not wasted or left unused and again that there is mutual non-interference. These principles are at present far from effective implementation, and doubts have been expressed whether the WARC on space communications has solved these problems in a satisfactory manner.

Proposals to be presented to the Conference have been examined by European broadcasters individually and in EBU. EBU has formulated a request for frequencies for broadcasting additional to the frequencies for the existing national services. Support is being sought for these proposals from all national delegations, and Unesco and the Council of Europe are interesting themselves. The proposition of the EBU, for example, would permit four to six additional direct broadcasting channels for reception in a European country. The EBU concept is based on the use of FM modulation and of narrow beam widths in certain frequency bands, which would make it possible to confine transmissions to a defined geographical area and to use the same frequency for a number of such limited services subject to a certain geographical separation. The importance in programme terms of this concept is obvious: the possibilities of national control, of international agreement on usage, of collaboration for common programme interests are immensely enhanced.

The question of adaptation of the ITU Convention and Radio Regulations is related to that considered under Question 3, namely, what is the best forum in which to establish international supervision or control of direct-satellite broadcasting? The problems have two parts, technical and political. The technical problems arise over the allocation of orbital 'slots' to satellites, and of radio frequencies to satellite broadcasting uses. The political problem arises over the choice and control of the content of broadcast programmes. The first would be best resolved by creating in the ITU a body, with sufficient technical resources and authority, to make allocations of orbital 'slots' and frequencies and so assume part of the functions of the existing IFRB. The second problem would be best resolved regionally by broadcasters' councils.

VI REPLY OF THE UNITED STATES GROUP

PRELIMINARY OBSERVATIONS

The general international legal setting for space communications and satellite broadcasting is established, in a constitutional sense, by the UN Charter and the Outer Space Treaty. The Charter and international law in general are expressly made applicable by the Treaty to outer-space activities. In addition, the Treaty contains specific provisions with a bearing on space communications. Article I prescribes free use of outer space by all states without discrimination and on a basis of equality of access. Article II prohibits national appropriation of outer space or any portion of it. Under Article VI, states bear international responsibility for national activities in outer space, whether by governmental or non-governmental bodies, and under Article VII, states are liable internationally for damage caused by objects they have launched. Jurisdiction and control over launched objects is in the state of registry, and ownership is unaffected by the space journey (Art. VIII). Under Article IX, parties to the Treaty undertake to carry out space activities 'with due regard to the corresponding interests of all others' and are obligated to consult in advance if there is reason to believe that a proposed activity 'would cause potentially harmful interference with activities of other States Parties'. Although these provisions are phrased in very general terms, their relevance to many phases of satellite communications is apparent.

At a more limited level, the provisions of other international legal instruments form part of the legal setting for states that have adhered to them. Chief among these are the ITU Conventions and Regulations, various copyright and related conventions, the Intelsat agreements, and, perhaps, the Intersputnik agreement. These instruments do not have the same constitutive significance as the Charter and Outer Space Treaty, and indeed one of the important issues raised by the Questionnaire is whether they or some of them should not be significantly modified or supplemented.

Finally, it has been suggested that a number of UN resolutions and declarations also contribute to the existing international law framework for space broadcasting. These documents may have significance as evidence of the law in particular concrete situations and may contribute to the efforts of judges and jurists to formulate the law. But they cannot be treated in themselves as authoritative statements of binding law. The weight and significance to be accorded them depends on the concrete issue and setting involved.

Against this general legal background, the US response to the IBI Questionnaire concentrates on situations where the problems, the application of principles, and the solutions can be concretely foreseen. More particularly, the US answers address the issues as they arise in a setting of a number of separate systems broadcasting to community antennas or receivers rather than to home television sets. Systems of the community type are already approaching the operational stage, and any necessary legal accommodation must be developed soon if it is to be timely. Moreover, both the Introduction to this IBI study and the UN Working Group agree that all actual systems put into operation for at least the next decade or more will be of this type.

The legal problems, particularly those involving propaganda, copyright, and other aspects of program content, are much simpler for systems where the remote broadcaster must pass through a gateway affording opportunity for control and surveillance than for systems with direct access to the private viewer.

If satellite broadcasting to community antennas can be freed for rapid development, the experience generated will be valuable in solving problems related to the home broadcast mode, if and when that mode reaches the operational stage. The alternative is not to suppress or hold back the development of community systems. These are already going forward and will continue to do so, whether or not a suitable legal environment is devised in advance. If present legal planning bogs down because of inability to agree on distant and possibly hypothetical issues essentially related to satellite broadcasting to home receivers, the result will be improvised or ad hoc solutions to legal problems as they arise in the community broadcast setting. Such solutions are likely to represent a less satisfactory accommodation of conflicting interests than could be achieved by timely and systematic analysis.

To concentrate on relatively definite, near-term problems is not to ignore the farther future. On the contrary, one important purpose of narrowing the present focus is to avoid foreclosing desirable lines of development on the basis of abstract considerations, before the exact contours of the problems have become clear and second-order consequences are better understood. To this end, a principle desideratum of any current legal arrangement is that it be to some extent open-ended and that it contain built-in mechanisms to permit ready adaptation to changing needs and technological developments. This is particularly important in the international legal system, which contains no generally operative judicial or legislative institutions for supple adaptation of the law to changing context.

QUESTION 1

What means of promoting and extending freedom of communication and information does direct-satellite broadcasting provide, in particular for developing countries?

REPLY

Every international discussion of satellite communications has stressed the potential of this new technology for promoting free access to information among peoples all over the world thereby improving understanding and interchange among them. At its second session the UN Working Group recognized

> ... that the development of technology for direct broadcasting from satellites holds the promise of unprecedented progress in communications and understanding between peoples and cultures. ... The Working Group considers that the medium of television is especially suited to increasing contacts between peoples of the world and to advancing the principles and purposes of the United Nations. Among the potential health benefits from direct broadcasts from satellites would be improved education and greater flow of news and information of general interest, including cultural programmes and the development of closer ties between peoples of countries and within countries.

Satellite broadcasting has a major role to play in implementing these objectives. And, as noted in the Introduction, the impact of this medium will be felt primarily in developing countries.

The succeeding sections of this paper deal with a number of the international legal problems that must be solved to bring this great potential to fruition. It is already apparent, however, that the critical constraints do not lie in the legal area. They arise because the developing countries are by and large without the technical and economic resources to launch, operate, and prepare programs for broadcast satellites. Unless these resources are made available in some form, the contribution of broadcast satellites to the growth and development of these countries will be another promise deferred. The main needs, apart from funds, which will always be in short supply, are in three fields:

Launch services: Only the United States and the USSR now have the ability to launch satellites. Japan and perhaps one or two other states or groupings of states may join this club within the next few years. The developing countries, and indeed all other states, will be dependent on these few sources for launch services for the foreseeable future. It has already been suggested that some sort of international agency be formed to provide launch services to all applicants on a kind of public utility basis. Such an agency might seek to develop its own launch capacity, which, as the European experience with

ELDO teaches, will be a long, painful, and difficult process. Alternatively, some or all of the existing launch states would have to put their facilities at the disposal of the international agency, but this seems an unlikely prospect. Short of an international launch facility, it would be desirable to develop a consensus on impartial criteria for the provision of launch services by the existing launch states. These criteria might ultimately be embodied in international agreement, but in the beginning they should probably take the form of a statement of guidelines or general principles. A second stage might be incorporation in a General Assembly resolution. The principles should prohibit the application of political conditions to the furnishing of launch services. In addition, the principles should ensure that direct-broadcast projects are soundly conceived in terms of system design and spectrum/orbit optimization. In the case of regional systems, the principles might include, in addition to technical characteristics, some consideration of the fairness of the terms on which access to the system is available to countries in the region. If, as is suggested below, there is some form of ITU plan for use of spectrum and orbital space, only systems that are not inconsistent with such a plan should be eligible for launch facilities under the agreed principles.

System design: Without experts of their own, the developing countries are substantially at the mercy of manufacturers from the developed countries for the conception and design of their communications-satellite systems. A number of manufacturers, often from different countries, are now in this field, and competition among them may provide some protection for the purchasing entities. It is at least as possible, however, that, as in the field of air transport, high-pressure salesmanship from competing suppliers may tend to push the developing countries into costly, ill-conceived programs that do not serve their needs well.

The principal requirement here is for an independent source of technical competence to which the developing countries can turn for analysis and advice. To some extent, this may be supplied by private consultants, but even so, a backstop is needed. The logical choice is the staff of the ITU, which already provides technical assistance in connection with UNDP communications programs. The technical capability of ITU will need to be both expanded and improved if it is to serve in an effective general advisory capacity on satellite communications for developing countries.

Similarly, national and multilateral development-financing agencies should scrutinize satellite broadcasting proposals carefully against both development and planning criteria. There is some basis for a priority in principle for communications projects. They contribute to the infrastructure of economic development, and, as noted in the Introduction, broadcast-satellite systems can have an enormous positive impact on literacy, occupational training, and the like. As important as the emphasis on sound telecommunications programs is a critical screening to avoid unsound and wasteful ones. Each proposal must be judged on its own merits in terms of the design and purposes of the particular system proposed, the overall priorities of the country involved, and integration with other aspects of a national development effort. If the ITU were to achieve a marked improvement in staff competence, it might serve as a point of referral for review and analysis of communications projects on behalf of potential financing agencies.

Technical assistance on programming: India, with its own extensive motion-picture industry, is exceptional among developing countries in being able to plan for largely indigenous program production. Elsewhere outside assistance will generally be needed, both in the actual production of programs and in the development of local production capability within developing countries.

As to the first, especially in the education field, much of the work in learning theory and in the use of visual methods for educational purposes has been done in the developed countries. This must be placed at the disposal of the new nations. Special care must be taken to anticipate cross-cultural differences and frictions. At first sight, it would appear that Unesco has an especially important role to play in this connection, but again there is the question of the availability of resources and technical competence. Regional broadcasting unions may be able to make a significant contribution as well.

Indigenous production is one field in which the developing countries should be able to move ahead rapidly. Costs of TV cameras and videotape should not be especially heavy, even by the stringent economic criteria that are applicable. Development and training of adequate production staff may be more difficult and time-consuming. However, the costly production values associated, for example, with prime-time commercial offerings in the United States, are unnecessary, and even out of place, in the setting of an educational and cultural program in a developing country. A modest application of financial resources and technical assistance should permit an increasing volume of production for these purposes to be performed within the country or region maintaining the system, in response to the needs of its own market.

A more general consideration bearing on freedom of information is that much of the discussion of legal rules for satellite broadcasting is concerned with restraints and inhibitions on programming, based on political or commercial grounds. Most of the remaining questions in the IBI Questionnaire reflect such preoccupations. A legal regime in this field must protect the legitimate interests of states and private persons. Without such assurance, the necessary resources of money and talent will not be forthcoming, and the potential of the broadcast medium will lie fallow.

But in focusing on particular public and private interests to be protected, the tendency is often to proliferate restrictions and close loopholes. There is hardly any form of expression that is not politically or culturally or religiously or personally offensive to someone somewhere. There is hardly any program transmission that cannot be made the occasion for seeking additional returns for author, producer, or performers. If the main effort is to prevent such expressions or exploit such opportunities, the result will be a legal strait-jacket, with systems operating at a bland and colorless universal least common denominator. The overall objective of promoting freedom of expression is lost from view.

The main thing may well be the policy perspective or predisposition from which legal proposals are regarded. The premise of the United States response to the remaining questions is that proposals for restraints are to be strictly scrutinized and are to be rejected unless there is a very strong showing that they are necessary to the proper development of the medium.

The prime objective of a legal regime for satellite broadcasting should be to vindicate and extend the market-place of ideas by ensuring the freest possible access to the medium, not only for listeners and viewers, but for senders as well.

QUESTION 2

In what ways, institutional or normative, may (a) hostile propaganda, (b) programs tending to disturb or erode religious, cultural, or social life, (c) advertising, be limited or prevented in satellite broadcasting consistent with the principle of freedom both to impart and receive information?

REPLY

The questionnaire identifies the three kinds of potentially undesirable programming that have chiefly been considered in discussions of satellite broadcasting: hostile propaganda, programs offensive on religious or cultural grounds, and commercial advertising. Lurking beneath the surface of these discussions, however, is the desire of many governments to be able to control the news and information reaching their people.

The first approach to the problem of unwanted or offensive programs, in the UN Working Group as well as other forums, was a proposal for a detailed internationally agreed code of program content, to which all broadcasting from satellites would have to conform.

The negotiation of such a code and its ratification on any wide scale would present almost insuperable difficulties, illustrated in the endless and fruitless efforts of UN bodies to define 'hostile propaganda' in other contexts. Too often, one country's news is another's hostile propaganda. Add the problem of defining what material may be offensive somewhere on religious or cultural grounds, and the difficulties of negotiation and draftsmanship increase at an exponential rate. It may be doubted that anything would be left in the permitted class. Even sports are not altogether immune, as is illustrated by the now well-known example of the bull-fight broadcast in India.

For the United States, and perhaps other countries with constitutional provisions guaranteeing freedom of speech, the code-of-content approach raises further difficulties. The First Amendment to the US Constitution provides that 'Congress shall make no law ... abridging the freedom of speech, or of the press. ...' Although the text of the Amendment refers only to Congress, the prohibition is interpreted as broadly applicable to all governmental agencies, local, state, and federal. One Supreme Court case, now some twenty years old, upheld a state law prohibiting 'group libel', *Beauharnais* v. *Illinois*, 343 U.S. 250 (1952), but that judgment is almost certainly undermined by later decisions of the Court. The position then is that except for the rapidly shrinking area of 'obscenity' and the general protection against defamation, invasion of privacy and fraud, there can be no government control of the content of expression by way of prepublication censorship.

Broadcasting is assimilated both to speech and the press within the meaning of the First Amendment, but the special characteristics of the broadcast medium have led to some limited regulation of program content. Radio and television stations in the United States are operated by private owners under license granted by the FCC 'if the public convenience, interest, or necessity will be served thereby'. Under this authority, the Commission requires that broadcasters allot some portion of their time to 'public service' programs. The requirement has not been given specific content in terms of number of hours or kinds of programs. It is usually satisfied by broadcasting of ordinary news shows, with perhaps some special events or discussion programs, usually in less desirable hours.

In the political field, the quasi monopoly position of the broadcaster has given rise to certain obligations. If he gives free time to one candidate for office, he must give equal time to all other candidates for the same office, both of major and minor parties. The proliferation of minor parties has provided a rationalization for not giving free time to any candidates except in very unusual circumstances. In the 1960 presidential election, the equal-time requirement was legislatively suspended so that the Kennedy–Nixon debates could be held. Except for the emanations of the obligation to serve the public interest, the broadcaster is under no specific duty even to *sell* time to political candidates or indeed to carry any political programming at all. If he does present one point of view on a matter of public controversy, he is obligated under a so-called 'fairness' requirement to give fair opportunity for the presentation of competing points of view.

The FCC has construed its licensing function to include authority to review program performance. It read the prohibition of censorship as limited to 'prior restraint'. The Commission could not suppress the broadcast of a particular program, but it could appropriately judge the overall program service of a licensee. The exercise of such responsibility might raise constitutional or statutory questions depending on the specific manner and extent of commission intervention. None the less, the agency concluded that it had to take the risk if it were to fulfill its mandate to grant original licenses, renewals, and transfers in the public interest.

In this view, the FCC has been upheld by the courts in a line of precedents, extending from 1930. The judiciary affirmed that the commission's authority extended over the 'scope, character and quality' of broadcast service, and that the agency's inquiry into a licensee's past program performance was not tantamount to censorship and suppression of free speech. In *National Broadcasting Co.* v. *United States*, 319 U.S. 190 (1943), the Supreme Court specifically rejected the contention that the Commission was limited to technical, traffic-control duties. In a famous phrase, Justice Frankfurter asserted that the Commission had 'the burden of determining the composition of that traffic'.

More recently, in *Red Lion Broadcasting Co* v. *FCC*, 395 U.S. 367 (1969), the Supreme Court confronted directly the reconciliation of First Amendment guarantees with the peculiar circumstances of a limited radio spectrum, and it clarified most of the ambiguity. The Court rejected the broadcasters' contention that the First Amendment gives them the right to use their allotted frequencies to broadcast whatever they choose, and to exclude whomever they wish from using that frequency.

Although broadcasting is clearly a medium affected by a First Amendment interest, *United States* v. *Paramount Pictures, Inc.* 334 U.S. 131, 166 (1948), differences in the characteristics of new media justify differences in the First Amendment standards applied to them. . . .

Just as the Government may limit the use of sound amplifying equipment potentially so noisy that it drowns out civilized private speech, so may the Government limit the use of broadcast equipment. The right of free speech of a broadcaster, the user of a sound track, or any other individual does **not** embrace a right to snuff out the free speech of others. . . .

Where there are substantially more individuals who want to broadcast than there are frequencies to allocate, it is idle to posit an unabridgeable First Amendment right to broadcast comparable to the right of every individual to speak, write, or publish. . . .

Because of the scarcity of radio frequencies, the Government is permitted to put restraints on licensees in favor of others whose views should be expressed on this unique medium. But the people as a whole retain their interest in free speech by radio and their collective right to have the medium function consistently with the ends and purposes of the First Amendment. . . . It is the right of the public to receive suitable access to social, political, esthetic, moral, and other ideas and experiences which is crucial here. . . .

In 1963 the FCC attempted to place a ceiling on advertising. It proposed a rule to limit commercial content according to the standards of the industry's own codes. Broadcasters enlisted the aggressive championship of the then chairman of the House Interstate and Foreign Commerce Committee. The House overwhelmingly supported his resolution denying the FCC's authority to issue rules limiting advertising. Although the FCC stoutly defended its legal prerogative, it withdrew its proposal with a statement that it had insufficient information to develop adequate standards.

The foregoing summarizes the total of government regulation of program content in US broadcasting, apart from the general legal prohibitions against obscenity, defamation, invasion of privacy, and fraud. It does not add up to much, but it probably goes to the constitutionally permissible limit of government power, at least in normal times. Moreover, in practice, the FCC has not defined more precisely the performance required under these general standards, and has seldom, if ever, refused to renew a station license for failure to comply with them. It will be seen that this constitutional position is in irreconcilable conflict with the notion of a code of content enshrined in law with its provision enforced upon broadcasters by the government.

The conflict between codes of content and the free speech guarantee of the US Constitution highlights a more general objection to this approach on the footing of policy. A commitment to freedom of information and the individual's right to know, enshrined in a number of international texts and resolutions, implies an orientation against control by governments of what their citizens hear, read, and see. Given the nature of broadcasting and the forms in which it is generally organized throughout the world, it will never be possible to eliminate a fairly large degree of government authority over content. The vice of the code-of-content approach, however, is that it brings governments in at the very outset to define the limits of what is permissible, and it establishes a governmental authority and duty to enforce those limits as a matter of international obligation. It thus engages governments fundamentally and pervasively in the business of regulating broadcast expression. To that extent it should be seen as inconsistent with the various internationally endorsed formulations of the principle of freedom of information.

As the deliberations of the UN Working Group have progressed, a growing consensus has seemed to be swinging against an international code of content for satellite broadcasting. It appeared at the third session of the Working Group that the principal remaining support for the idea came from East European countries, France, and certain developing countries. Discussion has turned instead to a more general principle that would prohibit satellite broadcasting to the territory of another country without that country's consent. The idea has had a number of formulations, none clearly defined in terms of practical mechanics. The two chief variants are one that would require the broadcasting state to obtain prior consent of all states where its programs might be received; and a second that would place the burden of objection on the receiving state.

The argument for this kind of restriction is by no means self-evident. The world has, after all, lived with a relatively open regime in the field of radio broadcasting for decades. Although there have been difficulties and frictions, there have been no unmanageable problems. Despite considerable efforts from a number of quarters, no government has been toppled by radio, and no wars have been won by it.

The Introduction concludes that there is little reason to anticipate that any country will establish a broadcast satellite with transmissions focussed outside its own territory for propaganda purposes, although the future use of such a system for commercial purposes may appear somewhat less unlikely. In the concrete technological and organizational context of a set of satellite systems broadcasting to community antennas, the only target for such a system would be community antennas in the receiving country. These would only be in place, however, if a national or regional system was already in operation for that country, presumably with frequencies and area of coverage registered with the ITU. For the sending system to broadcast into the same area on the same frequencies would in all probability create harmful interference in violation of the ITU Convention, except in the unlikely case that transmissions were confined to the periods when the national or regional system was off the air. In addition to any sanctions under ITU Conventions and Regulations, the state of the sending system would 'bear international responsibility' for the system's activities and would be 'internationally liable for damage', under the Outer Space Treaty. If the sender were undeterred by these obligations and consequences, it is unlikely that a further and more specific prohibition would be effective. But the final and decisive point is that even if an intruding satellite were to be emplaced, the objecting government, through its control over the community antennas, would still have the power to prevent reception. The most that could be achieved by the intruder would be to keep all programs off the air in the receiving state—hardly a realistic commercial or political objective.

In a regional or multilateral system, transmission from any member state would be able to reach community antennas in all states throughout the system. Here, too, however, there is no need for a special requirement of consent. The regional system would be established by agreement among the members expressly stipulating the terms and conditions of access to the system, the rules for sharing transmission capacity and whatever regulation of program content seem desirable. Such agreements for regional systems will probably be concluded among broadcasters after the manner of Eurovision, rather than among governments. They will vary from place to place and over time in accordance with the needs and experience of the parties. Again there seems little function for a universal regulation.

On the other hand, a requirement of consent in either of the forms now under discussion may seriously inhibit the development of community broadcasting. Despite increasing directionality of satellite antennas, it is still not possible to focus the transmission beam narrowly and accurately enough to confine it within the borders of a single country. A certain portion of the beam will spill over into areas not within the system. Even for large countries, the problem arises in border regions. It would perhaps be possible to devise some kind of *de minimis* exception to a consent requirement, but the definition is not easy.

Does a receiving state in these circumstances require the protection of a veto over its neighbor's system? If it has a satellite-broadcasting system of its own, it will necessarily be operating on different frequencies, since otherwise the spillover would create interference and would be impermissible under ITU rules. If the two systems are on separate frequencies, the antennas in the objecting state can be so constructed as not to be able to receive the spillover broadcast. If the objecting state has no broadcast system there would presumably be no community receivers to be reached by the spillover. Private parties or groups might be able to purchase a sufficiently sensitive receiver. Even if such receivers would cost no more than some hundred dollars, it is doubtful that many will be willing to make such an investment to pick up the unintended spillover from a foreign country. If so, and if the receiving government objects, the antennas would be visible and thus subject to its control. In this respect, community or augmented television receivers are in sharp contrast to ham radios, which can be cheaply made with locally available parts and readily secreted.

The fear about unwanted broadcasts is the fear that an outside broadcasting entity will have direct and uncontrolled access to viewers within the territory of a state. It has little or no reality in the context of community broadcasting. Whether the concern will have more substance for direct satellite-to-home broadcasting, if and when it arrives, will depend on a number of factors, including the technical and institutional forms such systems take. These cannot be foreseen now. It would be very unwise to constrict a community broadcasting capability that is within reach by a blanket-consent requirement relevant to a mode that will come into use, if at all, only after a decade or so.

Informal arrangements, such as guidelines or general principles, often promulgated by associations of broadcasters, have been useful in the past to help guide operators in the applications of the canons of good taste. Perhaps these associations might begin to address themselves to any special problems that may arise from satellite broadcasting. The declaration on the use of communication satellites adopted by the international conference of broadcast unions in March 1972 might be seen as a first step in this direction. For the present, the soundest course would be to impose no international legal restrictions, and to accumulate experience with community broadcasting. If some kind of consent requirement or other limitation seems desirable when satellite-to-home broadcasting is closer at hand and can be anticipated in more concrete detail, there will be time to impose it then.

At a minimum, any requirement of consent imposed at this time should operate only on broadcasts powerful enough to reach home receivers. It should be inapplicable to broadcasts that can only be picked up by community receivers. The line between these two categories is not bright and it is not defined by scientific criteria. It will require a judgment based on technical facts, which may have to be revised from time to time and will always have a certain arbitrary character. The function of establishing this dividing line would not appear to be beyond the political or technical capacities of the ITU. In any case, it is much better to face such difficulties of line drawing as there may be than to inhibit needed community-broadcast systems by giving each country a veto over any transmissions that might fall on its territory.

QUESTION 3

In what ways does the individual require protection in direct-satellite broadcasting from (a) defamation or false statements, in particular by a right of reply and the duty of rectification; (b) invasion of privacy, such as undue publicity of personal life?

REPLY

In the United States, defamation is governed by state law, subject to recently developed overriding limitations based on the First Amendment. That means that both the substantive definition of prohibited conduct and the remedies available to the successful plaintiff will vary from state to state in accordance with judicial precedents and legislative enactments in each. In general, the scope of the tort is rather narrowly drawn. Truth is a defense in almost every state even where malice of the defendant is shown. Malice and similar aggravating conduct of the defendant goes to damages.

The whole law of defamation is being reorientated in light of a recent Supreme Court decision in an action brought by a Southern sheriff against the *New York Times*, charging libel in the reporting of certain racial incidents in the sheriff's district, *New York Times Co.* v. *Sullivan*, 376 U.S. 254 (1964). The state court found for the plaintiff and imposed heavy damages. The Supreme Court reversed, holding that the First Amendment requires free and open discussion of the personality and conduct of public officials, uninhibited by the risk of prosecution for libel. Thus, publications with respect to such officials are immune from liability for defamation, even if the challenged statement is false and, perhaps, even if it is malicious, although this is not expressly decided by the case. The exact contours of this doctrine remain to be worked out, but it has been liberally construed and represents a fundamental change in the law of defamation. Since the rule is grounded in the Federal Constitution, it is applicable to all states uniformly, whatever their own basic law of defamation may be.

On the remedial side, the law in most states stresses damages. The strong bias, amounting in some states to a prohibition against injunctive relief, grows out of the same free-speech tradition discussed above. Traditionally there has been no right of reply in US defamation law, but in recent years it has begun to make its appearance in some states. An echo of the right is to be found, perhaps, in the 'fairness' doctrine, already referred to, requiring broadcasters, if they present one side of a controversial issue, to give fair opportunity for the expression of competing views. Newspapers and broadcasters, moreover, sometimes volunteer an opportunity for reply, and at a minimum this may bear on the issue of malice and the measure of damages.

Invasion of privacy is also a creature of state law in the United States so that no brief, definitive statement of its content is possible. The right of privacy is a late comer to the roster of interests protected by private law. Consideration of the subject dates from an article written by Justice Brandeis in the late nineteenth century, while he was still a student. Although Brandeis said later from the Supreme Court bench that the 'right to be let alone' was 'the most comprehensive of the rights of civilized man,' the states have not been hospitable in granting it recognition and have not been very successful in working out a systematic scheme of the interests involved. In the main, so far as is here relevant, the conduct protected against includes certain publications or discussion of persons or their characteristics that unfairly publicize their private or intimate lives or that unfairly exploit their public notoriety. Again, there is much more latitude in connection with public figures, who may be said to have opened their lives to public discussion, although this principle may vary with the facts of the particular case. On the other hand, public figures are ordinarily protected against commercial exploitation of their names and persons without their consent. Protection of the 'moral rights' of artists and authors, discussed below, frequently takes place under this rubric in the United States.

In this field, the courts may be readier to give injunctive or other specific relief than for defamation, but the principal remedy probably remains money damages. This would seem to make the whole experience even more painful for any truly sensitive person who thought his privacy was invaded. Perhaps this basic difficulty of protecting a right of privacy in a public judicial forum and by means of what is at best incommensurable relief accounts for the rather limited and erratic recognition given the right in the United States.

Because both defamation and invasion of privacy are governed primarily by state law, and vary from state to state, the United States experiences problems not unlike those faced on the international level in dealing with widespread publications. Thus, for example, a broadcast heard in several states may be defamatory according to the law of one and innocent or privileged according to the law of another. The relief available and particularly the measure of damages may vary. These conflicts are moderated to some extent because all the states are a part of the same legal system, but the differences remain significant. The response to these differences has not been an effort to develop uniform rules applicable throughout the country. Instead, the focus of scholarly and law-reform work has been on the question of appropriate rules of conflict of laws to govern the choice both of substantive law and remedies applicable to the various segments of a complex, and spatially-diffused situation. Despite much discussion and analysis along these lines, no very good solution has been worked out.

A main source of difficulty, from the analytic point of view, is the complex character of the harm involved in cases of defamation or invasion of privacy. There are two principal elements: the first is the offense to the individual and his sensibilities. This offense may vary somewhat with the place and extent of publication, but the gravamen of the harm is the publication of the offensive matter *per se*. The second element is damage to plaintiff's reputation, and this depends on the kind and quality of reputation he has in a particular community and the form and scope of the publication there. Although both elements are present in both torts, personal offense is ordinarily dominant in invasion of privacy cases and damage to reputation is the main ingredient of defamation.

The US experience suggests that an attempt to negotiate uniform international rules, whether of substantive law or conflicts of law, will not be very promising. The saving prospect again is that for the coming decade or more satellite broadcasting will be accomplished through the separate and relatively insulated national and regional systems using community antennas, rather than to homes all over the world from a single sending station. A national community-broadcast system will have spillover into neighboring states, but this is, in principle, no different than the spillover that occurs in terrestrial television broadcasting near national boundaries. Indeed, the problems may be smaller, because, as noted above, the neighboring state might have no community receivers at all, or if it did, they would point at a different satellite and most likely would operate on different frequencies than the receivers in the first state. In any case, the law of the forum, including its choice of law rules, would apply to the entire transaction, just as it does now in the case of national broadcasting entities.

In a regional system, it is possible that a broadcast originating in one member state would defame or invade the privacy of a person living in another and that the broadcast would be received in still other states participating in the system. If the states involved were of a single legal and cultural tradition, as for example in South America, it might be that the difficulties that have prevented the development of a uniform rule in the United States could be overcome. If so, any such rules could be embodied either in the basic agreement establishing the system or in an agreement among broadcasting entities governing content. If, as seems more likely, a consensus on substance fails, it should at least be possible to provide that broadcasters in any state within the system could be sued for defamation or invasion of privacy in any other member state by a plaintiff who is a national or resident of that state. This provision would give the plaintiff a convenient forum, and one that is not likely to be overly protective of the foreign broadcaster.

As regional and national systems develop direct satellite-to-satellite linkages, the area of reception of a single broadcast will expand. Neverthless, the principles outlined above should still be adequate.

The broadcast will not be received directly on the television screen from the remote broadcasting source, but rather mediately through the national or regional system. At least under US law, and probably elsewhere as well, that local system would be liable for any defamation or invasion of the right of privacy in the broadcast, at least to the extent of harm caused in the territory covered by the system. A solvent local defendant would be available to respond for damage to local plaintiffs, which should take care of most of the problem. Whether and to what extent the originating entity would have to indemnify the local system would be covered in the agreement providing for linkage between them.

This set of rules does not cover plaintiff's damages in states that are not part of his own national or regional system. Such damages may arise where the plaintiff has been harmed abroad by a broadcast not carried in his own state; or where the offending broadcast of foreign origination has been carried by systems in addition to that covering the state of the plaintiff. In such cases, the plaintiff is no worse off than under present law. If, for example, a citizen of the United States now believes himself defamed by a Eurovision program, his recourse is to sue the broadcaster in a European state in accordance with the law there prevailing. There is no compelling reason why he should be given additional rights because the broadcast is transmitted in Europe through satellite rather than terrestrial facilities.

It may be thought that linking separate satellite systems would result in much wider dissemination of objectionable material than has been likely up to now, and that the plaintiff is correspondingly more vulnerable. In many cases, the plaintiff's harm will decline with the distance from his domicile (perhaps in accordance with the rule of inverse squares), but not in all. The question whether protection should be provided in these situations is really one for the government and broadcasting authorities involved, to be decided in the light of their own appraisals of the affected interests. The matter could be covered in the agreement governing the terms and conditions of linkage between the systems. One could imagine, for example, a requirement that the originating broadcaster would respond to suit for world-wide damages in the state of which the plaintiff is a national or resident. This obligation would be considerably more burdensome than the parallel provision proposed above for regional systems. Most probably some safeguards against abuse would be needed, for example, a preliminary showing of probable cause by the plaintiff, a bond to cover some or all of defendant's costs if the suit is unsuccessful, or similar requirements.

QUESTION 4

In what ways, institutional or normative, do authors, performers and producers of broadcast programs, and producers of phonograms, need special protection for contributions to direct-satellite broadcasts in respect of (a) economic interests in the contribution? (b) the intellectual and artistic integrity of the contribution? Further, do satellite broadcasts themselves need special protection?

REPLY

(a) *Economic Interests in the Contribution*

International copyright protection today is provided under two major international agreements, the Berne Convention and the Universal Copyright Convention, and, to a lesser extent, by a number of bilateral and mutilateral conventions. Neither the Berne Convention nor the Universal Convention is adhered to by all major countries, but between them, they include almost all of the important countries with the exception of the USSR. Many of the signatories of the Berne Convention have also ratified the Universal Convention, but the United States and most Latin American countries are parties only to the Universal Convention.

The chief distinction between the two conventions is in the scope of the minimum protections that they provide. The Berne Convention imposes a high level of minimum protection and extends it over a broad range of the rights of the author, including not only the right to print and publish, but to translate and to authorize adaptations to other forms and media of communications. The Universal Convention also extends substantive protection to the right to reproduce, but beyond that only to the right of translation. In addition to the minima provided, both Conventions rely on the principle of national treatment to prevent discrimination against the foreign author. Under this principle, on matters not specifically provided for in the applicable Convention, the foreign work is entitled in any state bound by that Convention to the protection that is afforded a work originating in that state. An unfortunate consequence of the existing arrangements is that as between any two countries that are not members of the same Convention or parties to any bilateral or multilateral agreement, there is no copyright protection.

The principal problems posed by this international legal regime are its complexity and its spotty coverage. These defects cannot be corrected until there is more general agreement on appropriate minimum levels of protection. Still, as between developed countries parties to the same Convention, both produce fairly acceptable results. The Berne minima are higher, but the rule of national treatment in the Universal Convention provides rather comprehensive protection in most developed countries, which ordinarily provide a high standard of protection to their own works. There are annoyances due to variations in protection from country to country, but no encroachment on author's rights that could be deemed fundamental. A different situation arises as between a developed and a less developed country, where levels of national protection may be much lower. This issue is dealt with below in the response to Question 5.

As between developed countries, it does not appear that the advent of satellite broadcasting, particularly in the community mode, creates any significant new problems of international copyright or gives any additional urgency to the already felt need for harmonizing and universalizing the two existing sets of rules.

In a community-broadcast setting, the financial realization on copyrighted material included in broadcast programs would follow existing patterns, with only slight adaptations. The mechanisms need not and should not be uniform throughout the world, but would be adapted to those now in use in the particular area for terrestrial broadcast stations and networks. In the United States, for example, the producing organization or its selling agent is responsible for obtaining clearances on copyrighted material used in broadcasts. Under this arrangement, the networks perform the clearance function for the approximately 20 per cent of the programs they own. Motion-picture producers, who supply another immense slice of television programming, provide clearances for their product in the same way as for any other motion-picture exhibition. And independent producers must secure their own clearances. The charges are passed along to the broadcaster in one form or another as part of the cost of the program. The mechanics of this process can be much facilitated by associations of copyright holders, the oldest and most important of which in the United States is the American Society of Composers, Authors and Publishers (ASCAP). ASCAP issues to broadcast stations a blanket license permitting them to use all copyrighted music of members of the Society. The fee is fixed on the basis of the size and power of the station, the size of its audience, and the like. The revenues are pooled and, after deducting the cost of the organization, are divided among ASCAP members in proportion to the popularity of the works in general rather than the number of performances on a particular licensed station. The arrangements are endlessly complicated, but they are well understood among members of the broadcasting industry and their lawyers. No doubt similar arrangements exist in most other developed countries, indeed in rather simpler form, because the organization of broadcasting is more centralized.

These arrangements should be readily adaptable to a regime of numerous separate satellite-broadcast systems. National systems create no international copyright problems. Any spillover, as noted above, will not ordinarily be received by adjacent systems. In regional systems, the agreement establishing the system or an operating agreement among broadcasters would provide for clearance, probably by allocating responsibility to the originating state. The members of regional systems may well have similar copyright laws, philosophies, and clearance mechanisms. There may be, as in Europe, a body of experience with regional clearance for terrestrial broadcasts; or, as in Latin America, a special multilateral convention may be applicable. These special factors will affect and probably facilitate any arrangement for clearance of satellite broadcasts throughout the regional system.

In a wider network of national and regional systems, the terms and conditions of linkage will be agreed between the systems in advance, and the agreement should cover responsibility for copyright clearance. The choice is probably between placing responsibility upon the originating broadcaster to get clearances for every country in which the broadcast will be seen or upon each local system to obtain clearances in its own territory. The originating producer can know in advance what copyrighted material will be in the broadcast, which argues for placing the clearance responsibility on him. On the other hand, it will be a heavy burden if the program is to be widely shown, and he may be unfamiliar with local formalities and procedures. A central clearing house of some kind would be helpful here, though the problems of organizing one are formidable. A modest venture towards an information exchange on copyrights, primarily oriented to the needs of the developing countries, and without specific clearance responsibilities, has been set up by Unesco Perhaps it could ultimately be expanded to deal with this problem. Until then, perhaps the clearance burden could be shared, placing it on the local system for material that the originating broadcaster notifies in advance will be in the program, and otherwise on the originator. Agreements within or between separate systems would also provide for the allocation of royalty fees or other charges among the participants.

Satellite broadcasting, even in the community mode, presents the possibility of unauthorized reception and distribution of broadcasts. A sufficiently sensitive receiving antenna placed in an area where the satellite is visible could pick up the broadcast signals and relay the programs to receivers not within the sending system. If the programs carried copyrighted material that was protected in the area covered by the pirated broadcast, the broadcast would involve a violation of copyright, certainly by the pirate broadcaster, and perhaps by the system from which the program was taken.

As to the first, there is a similar possibility of international pirating today at least in areas close to national borders, but it seems not to have caused practical problems. Even within a single nation, there may be unauthorized pick-up and redistribution of a broadcast in remote areas far from the originating station's service area. In the United States, cable operators erected high antennas that could receive distant transmissions of popular network programs, often containing copyright material. These were then passed on by cable to the cable system's subscribers. (The right to *rebroadcast* such a program over the air is prohibited in the United States by the Federal Communications Act.) The cable companies paid no royalty fee on these distributions. When the issue was judicially tested, the position of the cable operator was sustained. It was held that the copyright monopoly was exhausted when the original permission to broadcast was given. Copyright owners argue that the rule is unfair because the original broadcaster will not be willing to pay copyright charges based on the remote audience; at the same time the cable distribution forecloses the possibility of earning the royalty through the sale of the program to a local broadcaster. The defenders of the cable position argue that a contrary view would subject the new medium to the domination of the networks and motion-picture producers, who control the copyright on the great bulk of US programming. The issues are now being considered in proceedings before the FCC, and in legislation pending in Congress. There is little prospect of an early decision in either forum.

This account of the US cable controversy is included as an illustration of the possibility of unauthorized redistribution of satellite broadcasts, not because it suggests the solution. A pirate broadcast would seem to call for application of the normal principles of international copyright law. If a copyright convention or treaty is applicable and the copyright holder is entitled to protection under it, he could presumably obtain injunctive relief and damages from the pirating entity in the courts of the state in which it is located. Ordinarily, as between developed states bound by the same international agreement, protection would be granted, although there could be anomolous situations as the account of the US cable experience shows. (Again the situation in the developing countries is different and is dealt with below.) Relief as against the original broadcasting entity would depend on the terms of the clearance that had been obtained from the copyright holder. If the clearance was geographically limited, the pirate broadcast would presumably be a violation. But it is not likely that a broadcaster would accept a clearance for an area narrower than that covered by his beam. By the same token, he will be unwilling to pay charges based on a theoretical audience within that area that his system does not reach

E

in practice. The balance between these considerations, however, can be left to negotiation between the interested parties—the copyright holders or their representatives, on the one hand, and the broadcasters on the other. There is no reason to suppose at this point that practical likelihood of pirating is great enough to prevent a negotiated settlement or to impose costs large enough to inhibit the growth of satellite broadcasting in developed areas.

The basic problems of dealing with international pirating of copyrighted material, whether in broadcast or other media, remain. They are to reach international consensus on the scope of protection and to get as many nations as possible to provide the agreed protection. These ends will be achieved more readily if all concerned retain a sense of proportion about their importance and are modest about the scope of protection, than if they relentlessly seize every occasion to extend it. The advent of satellite broadcasting does not add any really new problems, and it probably will not significantly complicate or magnify the existing ones.

(b) *The Intellectual and Artistic Integrity of the Contribution*

The Berne Convention protects the 'moral' rights, so-called, of the creator of a copyrightable work. Article 6 *bis* (1) (Rome text) provides: '... the author shall have the right to claim authorship of the work as well as the right to object to any distortion, mutilation or other modification of the said work which would be prejudicial to his honour or reputation'. The Universal Convention contains no similar provision. In the United States, the matter is in large part remitted to the states to be dealt with by each of them according to its own notions of policy, subject to the overriding limitations of the First Amendment. The position is very much like that prevailing with respect to defamation and invasion of privacy. In general, US courts are not very hospitable to these 'moral' rights, and it is unlikely that any very generous recognition would be permitted by the First Amendment. Such protection as is given is likely to focus on the economic interest of the author, rather than his artistic sensibility or reputation. It may be that the artist has some rights to prevent the physical alteration of his work even after it has left his hands, but this would seem to have little relevance to the broadcast medium. Perhaps, also, the true creator of a work would be able to prevent another from claiming authorship, on ordinary rules against passing off and misappropriation. In the famous case of *International News Service* v. *Associated Press*, 248 U.S. 215 (1918) a news service that had been denied access to a theater of war was prevented from pirating a competitor's uncopyrighted dispatches and distributing them to its own clients. But the case has not had a large progeny. A copyright holder may be able to do something to protect his 'moral' rights by exercising his power to prevent publication of excerpts or abridgements. (Such rights, unlike the others discussed, would derive from the Federal Copyright Act, rather than state law.)

It may be doubted whether as a matter of policy any very great indulgence is due the 'moral' rights of the artist. Intellectual life has always thrived on caricature, parody, lampoon, and satire, and an important ingredient of these genres has always been the 'distortion, mutilation or other modification' of works of art in ways that are not too careful about the author's 'honor or reputation'. Even serious criticism might fall within the Berne language, broadly construed. Freedom of expression would be ill served by a set of rules that significantly restricted these activities, with not much in the way of offsetting gain. Authors and artists flourish when they are free to create and others are equally free to comment on their creations.

Whether or not these arguments of policy are persuasive, the development of satellite broadcasting should not alter the balance. It provides no new means or occasion for offending the artist by distorted renditions of his works. The possibility that the distortions may reach a wider audience, after linkages between satellite systems become available, provides no independent ground for restriction.

(c) *Do Satellite Broadcasts Themselves Need Special Protection?*

The question here deals with the protection of the broadcast itself apart from the component literary, artistic, and musical material that can already get copyright protection. Broadcasts as such do not fall within either of the major copyright conventions, although a broadcast script may be copyrighted, just like a play, and a videotape of a broadcast may presumably be copyrighted and further showings of that tape or of copies of it may thus be controlled.

On the international plane, the subject of broadcasts as such is dealt with in two specialized instruments, neither of them having many adherents. The Rome Convention of 1961 grants to 'broadcasting organizations' the right to authorize or prohibit rebroadcasting of their broadcasts, fixation of the broadcasts (e.g. on tape), the reproduction of fixations, and distribution to places where the public pays a fee to enter. In addition the Convention calls for national treatment of broadcasting organizations and broadcasts from other signatory states. Only a dozen or so states have signed the Rome Convention, some with significant reservations. A European Agreement on the Protection of Television Broadcasts grants even more sweeping rights, but it is designed specifically for Europe.

The US Federal Communications Act makes it unlawful for any broadcasting station to 'rebroadcast the program or any part thereof of another broadcasting station without the express authority of the originating station.' The doctrine of unfair competition has been successfully used by radio and television broadcasters to prevent rebroadcast, or showing for a fee, of sports and news events. Since common-law copyright is sometimes questioned for broadcast of events in the public domain, unfair competition offers a convenient supplement to statutory copyright. Similarly, the doctrine of unfair competition is used to protect titles to literary works.

As noted above, the provision against rebroadcasting has been held inapplicable to the redistribution of a broadcast signal over wires to the subscribers of a cable system. The rule seems to have been adopted in view of the decentralized organization of the broadcast industry in the United States with its many privately-owned local stations, and with national networks providing common programming to affiliated stations. A relatively small number of other countries have adopted legislation bearing on the matter, often as part of a code regulating broadcasting rather than as a matter of copyright.

The lack of enthusiasm for international agreement and the dearth of national legislation suggests that, except in special circumstances, the unauthorized use of broadcasts has not created much of a problem. It may be inferred from the existence of a federal law in the United States and a widely applicable European Agreement that the special circumstances are the coexistence in an area of a large number of stations whose fields of transmission partially overlap. In the satellite-broadcast field, the prospect is for numerous national and regional systems, with directional but not accurately focussed beams confined to the area within the system. It may be argued that this situation will begin to approximate that in terrestrial broadcasting in the United States or Western Europe, so that similar protection of the broadcast, as such, is desirable.

This conclusion should be treated with great reserve. The Introduction concludes that there will be relatively few separate satellite systems. Most, if not all, will be government owned or controlled entities. It seems quite unlikely on the basis of terrestrial experience with both radio and television that these entities would make unauthorized use of each other's broadcasts. There is some possibility of pirate receiving and distribution stations, discussed above in connection with copyright, but there is simply no way to anticipate how large a problem this is likely to be.

More important, a great many, perhaps most of broadcast programs contain copyrighted material and are, therefore, in part covered by copyright. In that way the broadcaster can maintain control of the program.

Only programs containing no substantial components that could be protected by copyright would be significant beneficiaries of the proposed protection of broadcasts as such. What kinds of programs are these? Primarily, they would be live programs devoted to news, public affairs, panel discussions, and current events. *A priori*, there is little reason for subjecting such broadcasts to restrictions, except to provide the broadcaster with additional revenue. In the absence of a convincing showing that he needs such revenue in order to continue these programs or that the protection would provide an incentive for a large increase of programs in this category, the interest of the broadcaster in protection is outweighed by the public interest in widespread dissemination of such material. Sports events might be an exception to this generalization. They hardly seem an appropriate subject for a special international treaty, but this may be a field for an agreement among broadcasters. On the whole it may be preferable to preserve the principle that one cannot obtain monopoly rights in matters of general and current interest merely by transmitting images of those matters to the viewing public.

QUESTION 5

What problems do the issues raised in Question 4 create for developing countries in the reception of direct-satellite broadcasts or access to satellites?

REPLY

Television broadcasting is now a minor activity in most of the developing world. The area of coverage is small, and the audience is confined, in the main, to a narrow circle of the affluent and the urban middle class. Programs are generally inferior and often consist of out-of-date re-runs of material from the developed countries, perhaps with a sound track in the local language dubbed in.

All this will change with the introduction of satellite-broadcasting systems. Vast new audiences will be brought within reach of the screen, and for the first time, the principal purpose of the TV medium will be not entertainment but enlightenment. To achieve these ends, as has been noted above, it is of the highest importance that first-rate programs and program material be available to those systems at prices they can afford. Often that will be a nominal amount.

This prospect is viewed with alarm by copyright holders in developed countries; one wonders why. Copyright protection, after all, is not a part of the Ten Commandments; it is of fairly recent date. And it has not usually been favored by countries in the course of development whose own literary and artistic production was small. In the United States, the first comprehensive copyright law was in 1909, and until then, foreign works had no protection at all within its borders.

Discussion of copyright policy usually proceeds in an atmosphere dominated by the perception of the author's—in practice, often, the publisher's—property rights in the artistic creation. Less weight is given to the countervailing public interest in the free dissemination of these productions. As a result the tendency of copyright law, on both the national and international plane, has been steadily to increase the scope and duration of the copyright monopoly. It is questionable whether the present levels of protection can be convincingly justified as a needed incentive to creation.

The copyright provides a useful means of compensating authors and publishers (and their analogues in other media) in societies that can afford to pay for their works. But it ought not to be a device for denying access to those works to societies that cannot pay in any event. To do so gives no benefit to the author and trenches sharply on the interest in freedom of information. Moreover, in the final analysis, it is inimical to a regime of copyright law.

If the developing countries are denied access to needed material within the framework of international copyright law and administration, they will take it without benefit of that framework. The United States did so for the first century of its existence when it was a developing country. The only way to prevent large scale pirating of works in the developing countries is to make sure that the material is available at the nominal prices they can afford. If such arrangements are not made, and the new nations resort to self-help, the ultimate losers will be the copyright owners themselves, for the establishment of an orderly regime of universal protection for creative works will be that much longer deferred.

There seems to be a wide measure of agreement with these conclusions—whether from conviction or necessity—but indifferent success in carrying them into effect. The major effort to accommodate the copyright needs of developing countries has been through special provisions in the principal Conventions. The first of these was the Protocol to the Stockholm text of the Berne Convention. This Protocol provides that developing countries may decline to maintain the minimal levels of protection required by the text of the Convention with respect to duration of protection, copying, broadcasting, and educational uses. Only two countries have adhered to the Stockholm text to date, and it must be accounted a failure. The 1971 revision of the Berne Convention and the Universal Copyright Convention are partly designed to assist the developing countries. They permit developing countries that are parties to the Berne Convention to withdraw and adhere to the less stringent Universal Convention without the penalties now applicable in such a case.

There is considerable evidence that holders of copyrights in fact license them for use in developing countries at reduced or nominal rates. US television programming is sold in these countries at a flat rate of $25 per half hour, a figure which very probably does not cover selling costs. A study made by the United States in preparation for a meeting of the International Copyright Joint Study Group revealed comparable practices among book publishers.

Apparently the holders of secondary rights, the performers, do not take the same attitude. Each new performance may trigger compensation requirements under their rights, and the fee scales are often not realistically related to the economic situation in developing countries.

It may be that this scattered evidence indicates a more fruitful avenue of approach than formal amendment of international treaties. It might be possible, perhaps under the auspices of Unesco, to secure agreement from the principal organized associations of copyright holders—publishers, producers of broadcasts and motion pictures, composers, and performing artists—that licenses would be given at nominal rates for use in developing countries, at least for news, educational and cultural purposes broadly defined. This approach would maintain the integrity of the existing system of international copyright protection, and would not deprive the copyright holders of any significant revenues, since licenses would only be issued for areas where no significant commercial market exists. The overall agreements would give visibility and formality to what is probably widespread practice in any event. And agreements with private associations might be more flexible and easier to negotiate than treaties among sovereign states.

A blanket nominal royalty license would have the further advantage of simplifying the clearance procedure, which is now exceedingly burdensome. For educational television in the United States it is said that clearance requires six months lead time, and that two-thirds of the total amount spent represents the administrative costs of obtaining clearance, with only one-third going to actual royalty payments. The problems of obtaining clearances through normal procedures would be even more difficult for developing countries. The blanket-license technique would obviate much of the difficulty.

In addition, Unesco has established an international copyright information center. According to the original proposal, the center was to 'be established initially on a modest scale', which means principally that it should not cost much money and that it should not perform the service of actively clearing rights. It may be that this 'modest' effort is all that is possible at the present time, but surely it should go forward promptly and should be expanded as rapidly as the occasion permits.

Finally, the needs of developing countries are another reason for refusing to accord monopoly in broadcasts as such. In the whole field of artistic and intellectual property, an overriding need is to meet the requirements of developing countries by finding a way to reduce existing restrictions. This has not proved an altogether easy task. It makes little sense to create a new set of monopoly rights, only to be faced with the difficult technical problem of devising appropriate exceptions.

QUESTION 6

How is direct satellite-broadcasting to be correlated with other systems of satellite communications and in particular (i) what problems does direct-satellite broadcasting create in respect of the determination of orbital positions of satellites, the allocations of frequencies, measures to avoid interference with other transmissions, and measures to limit or prevent spillover; (ii) what adaptations are necessary to deal with these problems in the ITU Convention and Radio Regulations?

REPLY

1. *United States Regulatory Setting*

The basic framework for US government regulation of the spectrum is established by the Communications Act of 1934. The Act (as amended) divides spectrum-management responsibility between the President and an independent regulatory commission, the FCC. The President is given the authority to assign frequencies to government agencies, a task now formally delegated to the director of the Office of Telecommunications Policy (OTP). The OTP, a branch of the Executive Office of the President, is in turn advised by an Interdepartmental Radio Advisory Committee (IRAC) of user agencies. The FCC is not a member of the IRAC, but sends a representative to maintain liaison.

Together with the FCC, the OTP arrives at a National Table of Frequency Allocations which establishes the division of the spectrum between government and private use. Within this latter category, the FCC has the responsibility to allocate frequencies among classes of users, to license and assign frequencies to particular users, and to establish rules to prevent harmful interference. In exercising its assignment and allocation powers, the FCC is subject to the requirements of procedural due process outlined in the Administrative Procedures Act, and to the substantive standard of the Communications Act that licenses be granted according to the 'public interest, convenience and necessity'. No license may be granted for a period exceeding five years, but the Commission may renew licenses for the same period as the original term.

By formal rule-making proceedings, the FCC has established a table of allocations (similar to that of the ITU) reserving portions of the spectrum for particular types of users. Commission rules also specify restrictions on the power, location, frequency, and type of modulation of transmitters used by licensees, in order to preclude interference with other users.

Prospective users of the spectrum who propose a use that fits within the table of allocations, conforms to the engineering requirements of the FCC, and would not conflict with other current or proposed uses are routinely granted licenses by the Commission. Where the new use would conflict with an existing use, the FCC will either attempt to redesign the proposal to avoid conflict or will deny the application, since 'squatter's rights' are usually accorded to the established user. When two inconsistent applications are before the FCC, it may choose to resolve the conflict by a formal comparative hearing. This procedure is followed most commonly in the case of multiple applications for broadcast frequencies, but it has also been employed with respect to non-broadcast uses.

Assignments of frequencies to government users are made by the OTP by means of a similar, though less formal, procedure. The OTP'S Manual of Regulations and Procedures for Radio Frequency Management establishes technical standards, and serves the same function as the FCC's formal regulations. Most conflicts are resolved by the operating agencies or within IRAC, and do not require consideration by the OTP.

The position to be taken by the American government in international radio negotiations is arrived at through deliberations within the IRAC, and may be placed formally before the public for comment by the mechanism of an FCC Notice of Inquiry. The Spectrum Planning Subcommittee of IRAC had the major responsibility for preparing the US position paper for the 1971 WARC.

For satellite communications, the FCC authorized the use of only a part of the band allocated by the 1963 Geneva conference. Allocations are shared with terrestrial microwave stations. In addition to these frequency limitations, technical standards are imposed which limit the permissible power-flux density of satellite transmission reaching earth stations; conversely, terrestrial microwave stations are restricted to a power ceiling. Interference between terrestrial microwave stations and earth stations serving satellite telecommunications is minimized by requiring earth stations to be located in isolated areas, well away from population centers; and there are limiting requirements as to the angle of elevation of earth station antennas and as to the maximum effective radiated power of transmissions from satellites. According to an FCC publication,

> the limits for each service are well in excess of the values used today so that each has room for expansion in the power domain. [There is also] a mandatory coordination procedure to ensure that once an earth station is established, its capacity will not be expanded in such a way as to cause harmful interference to existing microwave systems and conversely, that

new microwave stations will not cause harmful interference to existing earth stations.

Domestic use of communications satellites has been proposed to the FCC by a number of applicants, but as yet no decision has been reached on the ownership or operation of such satellites. A policy paper issued through the White House in 1969 suggests that the FCC establish a policy of allowing any user or communications carrier to operate a domestic satellite system, and recent FCC decisions allowing greater competition among terrestrial carriers may signal a favorable disposition on the part of the Commission. But in the event of multiple applications for domestic-satellite earth stations, problems of technical interference might require the FCC to choose among applicants even if the Commission has no wish to do so on economic grounds. In the event of such inconsistent applications, it is not clear whether the FCC would resort to formal competitive hearings or to a more customary seriatim-application process.

The procedure for acting on proposals for frequency assignments for Intelsat satellites has been as follows: Intelsat provides information concerning frequency use to ITU members in accordance with resolution 9(A) of the 1963 radio conference. After member countries have had a chance to comment and any adjustments have been made, Intelsat forwards the information through the Communications Satellite Corporation (Comsat) to the FCC, which registers the frequency use with the ITU. The FCC does not at this stage exercise any control over the frequency assignment, but merely intercedes to satisfy the requirement that frequencies be registered only by member states. Proposals will contain specifications for frequency, power, and orbital placement, and hence will constrain the location of earth stations necessary for access to the satellite. After the proposal has been agreed upon by Intelsat and the FCC, the Commission will issue a construction permit for the space device, a grant to perform tests on the transmitting equipment, and finally a letter of authority to operate the transmitters aboard the satellite. Such a letter of authority does not have precise legal status, and is not issued for a fixed term. Transmitters in earth stations located in the United States are, however, licensed in the same way as terrestrial microwave transmitters.

2. *The International Setting*

Satellite communications make use of two international resources: the radio-frequency spectrum and space along the geostationary orbit in which to emplace the satellites. Both of these resources are, in a sense, finite. The Introduction suggests that these resources are *potentially* adequate for all conceivable future needs. But it is fairly apparent that existing legal and institutional arrangements do not provide assurance that this potential will be realized in practice.

The Radio Regulations do not prescribe a first-come-first-served rule, but this is the practical outcome of existing procedures, at least as between otherwise complying systems. It does not appear that the new procedures and recommendations adopted by the 1971 WARC will significantly alter this basic condition. It follows that there is no incentive for early registrants to adopt system designs that optimize or even economize the use of the spectrum orbit resources. These early systems may therefore unnecessarily preempt spectrum and orbit that would otherwise be available for late comers. The problem may be moderated somewhat by the relatively short life span of communications satellites, heretofore thought to be about five years. But this period is increasing, and in any case does not provide a positive control over future spectrum/orbit use.

In response to the inadequacies of the first-come-first-served principle, it has been suggested that a positive allocation of spectrum and orbit resources be affected through the machinery of the ITU. Presumably such an allocation would provide exclusive or priority rights for each member to certain frequencies at given orbital positions, based on standard technical specifications as to satellite spacing and other variables. Such a system would have the virtue of preserving some access to spectrum/orbit resource for each country until it was ready to go forward with its satellite-communication plans.

This kind of allocation scheme would have disadvantages of its own. In the first place, it is unduly rigid. Although some possibilities for adjustment could be built into the allocations, it is unlikely, especially given present ITU machinery, that sufficient flexibility could be achieved to adapt readily to the rapidly changing technological and economic setting in satellite communications. Moreover, advance allocation may result in unnecessary preemption of spectrum and orbit by non-users, or late comers, just as first-come-first-served may mean preemption by early users. Most important, a fixed allocation based on standard design specifications will not be able to achieve the full communications capacity of available spectrum and orbit. That will require full exploitation as between particular systems sharing a certain portion of the orbit, of the design variables available with the technology at any particular time.

What is required is not allocation on a rationing basis, but a flexible system of advance planning and coordination of technical characteristics of systems to ensure that each system incorporates available design options for minimizing spectrum/orbit use at acceptable cost. In addition, the planning activity should ensure the availability of resources for late comers, by reserving necessary space, by imposing time limits on earlier systems, and by developing a fair procedure for sharing the costs of necessary adaptations of earlier systems. To be effective, the planning and coordinating activity would have to be on a continuing basis.

As a practical matter it would appear that this planning function should be vested in the ITU. The ITU does not do this kind of planning at present, nor does it have the authority or the technical staff to be able to do it. It is improbable that the members of ITU would agree to any other assignment of this responsibility, however, and the function is sufficiently related to existing ITU activities to make that organization the logical place to put it.

Ideally, the product of the planning activity should be submitted periodically to a political organ of the ITU for review and approval, after which the provisions of the plan would assume the character of legally binding obligations on ITU members. Given the history and tradition of ITU, however, it is very unlikely that the members would agree to these legal consequences, much less establish machinery for their effective enforcement. And it may be that this kind of formal legal status would be unnecessary. If the planning activity is carried out skilfully, with careful consultation of affected parties at all times—particularly early in the design phase of proposed systems—compliance with the plan should be relatively easy and not very costly. There would be little or no incentive for violations, and for these, the ordinary international sanctions of censure, formal or informal, may be successful deterrents. In these circumstances, 'indicative planning' may be sufficient.

The ITU will have to be strengthened if it is to discharge this planning responsibility effectively. According to a proposal put forward by the Twentieth Century Fund Task Force, a permanent planning group should be established made up of engineers, communications specialists, economists, lawyers, and other relevant experts. These should not sit as national representatives, nor act under government instructions. The planning group would assemble an inventory of current and future plans for communications-satellite systems, both for broadcasting and other purposes. Members of the ITU would undertake to file their plans with the group. The group would seek to integrate these separate system proposals into a comprehensive plan, taking account of the variations between the three ITU regions in terms of service requirements and spectrum use. If conflicts appeared as between the proposals put forward by different systems, the planning group would be empowered to develop alternative design proposals to eliminate the conflict with the least cost or performance penalties. The plan would be reviewed and revised annually in the light of new filings and other developments.

The planning group would need the backing of a solid technical staff. This would require adequate salaries for high quality men and women and probably considerable consolidation of ITU technical personnel, who are now divided among three semi-autonomous staffs.

Whether this proposal or a more modest one is adopted, effective planning for broadcast satellite communications will require changes in the present powers and structure of the ITU.

APPENDIX OF DOCUMENTS

1. *UN General Assembly Resolution 2733—xxv:* International Co-operation in the peaceful uses of Outer Space.
2. a. Agreement relating to the International Telecommunications Satellite Organization 'Intelsat' (20 August 1971): Articles I (k)–(n), II–V.
   b. Intelsat operating Agreement (20 August 1971): Articles 13–18.
3. Declaration by the International Conference of Broadcasting Unions on Communication Satellites, Rome (11 March 1972).

The Intelsat Agreements replace the Intelsat Interim Agreement of 1964, and came into force upon acceptance by two-thirds of the States parties to the Interim Agreement, including parties which then held two-thirds of the quotas.

Chosen for reproduction here are the provisions of both the new Agreements, which represent the functions and operations of the Intelsat system.

I

2733 (XXV). INTERNATIONAL CO-OPERATION IN THE PEACEFUL USES OF OUTER SPACE

A

The General Assembly

Recalling its resolution 2453 B (XXIII) of 20 December 1968 whereby it established a Working Group of the Committee on the Peaceful Uses of Outer Space to study and report on the technical feasibility of communication by direct broadcast from satellites and the current and foreseeable developments in this field, as well as the implications of such developments in the social, cultural, legal and other areas.

Taking note with appreciation of the reports prepared by the Working Group on Direct Broadcast Satellites during its three sessions,[1]

Noting that the first satellite-borne instructional television experiment for direct reception into community receivers will be undertaken in India as early as 1973/1974, thereby making it possible to enrich life in isolated communities,

Noting that the potential benefits of satellite broadcasting have particular significance with regard to better understanding among peoples, the expansion of the flow of information and the wider dissemination of knowledge in the world, and the promotion of cultural exchanges,

[1] *Twenty-fourth Session, Supplement No. 21A* (A/7621/Add.1), annexes III and IV; and *ibid., Twenty-fifth Session, Supplement No. 20* (A/8020), paras, 48–59.

Recognizing that the use of satellite-borne television for educational and training purposes, particularly in developing countries, can in many instances contribute towards national programmes of integration and community development and economic, social and cultural development in such areas as formal and adult education, agriculture, health and family planning,

Taking note of the concern of the Committee on the Peaceful Uses of Outer Space in considering the practical interests of all States, in particular the interests of the developing countries, regarding the efficient use of geostationary orbit and the frequency spectrum,

Recognizing that the effective deployment and use of direct satellite broadcasting requires large-scale international and regional co-operation and that further consideration may have to be given to the legal principles applicable in this field,

Endorsing the Working Group's conclusions on the applicability to such broadcasting of certain existing international legal instruments, including the Charter of the United Nations, the Treaty on Principles Governing the Activities of States in the Exploration and Use of Outer Space, including the Moon and Other Celestial Bodies and the applicable provisions of the International Telecommunication Convention[1] and Radio Regulations,

1. *Recommends*, on the basis of the probable patterns of use of satellite broadcasting systems outlined by the Working Group on Direct Broadcast Satellites of the Committee on the Peaceful Uses of Outer Space, that Member States, regional and international organizations, including broadcasting associations, should promote and encourage international co-operation at regional and other levels in order, *inter alia*, to allow all participating parties to share in the establishment and operation of regiona satellite broadcasting services and/or in programme planning and production;

2. *Draws the attention* of Member States, specialized agencies and other interested international organizations to the potential benefits to be derived from direct broadcast satellite services, especially in developing countries, for improving their telecommunications infrastructure, thereby contributing to general economic and social development;

3. *Recommends*, with a view to making available the benefits of this new technology to countries, regardless of the degree of their social and economic development, that Member States, the United Nations Development Programme and other international agencies should promote international co-operation in this field in order to assist interested countries to develop the skills and techniques that may be necessary for its application;

4. *Requests* the Committee on the Peaceful Uses of Outer Space to keep under review the question of reconvening the Working Group on Direct Broadcast Satellites at such time as additional material of substance on which further useful studies might be based may have become available;

5. *Recommends* that the Committee on the Peaceful Uses of Outer Space should study through its Legal Sub-Committee, giving priority to the convention on liability, the work carried out by the Working Group on Direct Broadcast Satellites, under the item on the implications of space communications;

[1] Signed at Montreux on 12 November 1965.

6. *Invites* the International Telecommunication Union to continue to take the necessary steps to promote the use of satellite broadcasting services by Member States and to consider at the 1971 World Administrative Radio Conference for Space Telecommunications the appropriate provisions under which satellite broadcasting services may be established;

7. *Requests* the International Telecommunication Union to transmit, when available, to the Committee on the Peaceful Uses of Outer Space, all information about the use of the geostationary orbit and the frequency spectrum;

8. *Invites* the United Nations Educational, Scientific and Cultural Organization to continue to promote the use of satellite broadcasting for the advancement of education and training, science and culture and, in consultation with appropriate intergovernmental and non-governmental organizations and broadcasting associations, to direct its efforts towards the solution of problems falling within its mandate.

1932nd plenary meeting,
16 December 1970

B

The General Assembly,

Recognizing the importance of international co-operation in developing the rule of law in the exploration and peaceful uses of outer space,

Recalling that, in its resolutions 1963 (XVIII) of 13 December 1963, 2130 (XX) of 21 December 1965 and 2222 (XXI) of 19 December 1966, it requested the Committee on the Peaceful Uses of Outer Space to prepare a draft convention on liability for damage caused by objects launched into outer space,

Recalling that in its resolution 2345 (XXII) of 19 December 1967, in which it commended the Agreement on the Rescue of Astronauts, the Return of Astronauts and the Return of Objects Launched into Outer Space, it also called upon the Committee on the Peaceful Uses of Outer Space to complete urgently the draft convention on liability,

Recalling also its resolution 2453 B (XXIII) of 20 December 1968, in which it requested the Committee on the Peaceful Uses of Outer Space to complete urgently the draft convention on liability and to submit it to the General Assembly at its twenty-fourth session,

Recalling further its resolution 2601 B (XXIV) of 16 December 1969, in which it urged the Committee on the Peaceful Uses of Outer Space to complete the draft convention on liability in time for final consideration by the General Assembly during its twenty-fifth session and emphasized that the convention was intended to establish international rules and procedures concerning liability for damage caused by the launching of objects into outer space and to ensure, in particular, prompt and equitable compensation for damage,

Affirming that until an effective convention is concluded an unsatisfactory situation will exist in which the remedies for damage caused by space objects are inadequate for the needs of the nations and peoples of the world,

Aware that various proposals have been submitted to the Committee on the Peaceful Uses of Outer Space and that a number of provisions have been agreed upon, although subject to certain conditions and reservations, in its Legal Sub-Committee,

1. *Takes note* of the efforts made by the Committee on the Peaceful Uses of Outer Space and its Legal Sub-Committee at their sessions in 1970 to complete the preparation of a draft convention on liability,[1] for submission to the General Assembly at its current session;
2. *Expresses its deep regret* that, notwithstanding some progress towards this objective, the Committee on the Peaceful Uses of Outer Space has not yet been able to complete the drafting of a convention on liability, a subject which it has had under consideration for the past seven years;
3. *Affirms* that the early conclusion of an effective and generally acceptable convention on liability should remain the firm priority task of the Committee on the Peaceful Uses of Outer Space and urges the Committee to intensify its efforts to reach agreement;
4. *Notes* in this connexion that the main obstacle to agreement lies in differences of opinion within the Committee on the Peaceful Uses of Outer Space on two main issues: the legal rules to be applied for determining compensation payable to the victims of damage and the procedures for the settlement of claims;
5. *Expresses the view* that a condition of satisfactory convention on liability is that it should contain provisions which would ensure the payment of a full measure of compensation to victims and effective procedures which would lead to the prompt and equitable settlement of claims;
6. *Urges* the Committee on the Peaceful Uses of Outer Space to make a decisive effort to reach early agreement on texts embodying the principles outlined in paragraph 5 above with a view to submitting a draft convention on liability to the General Assembly at its twenty-sixth session.

1932nd plenary meeting,
16 December 1970.

C

The General Assembly,

Recalling its resolutions 2600 (XXIV) and 2601 (XXIV) of 16 December 1969,

Having considered the report of the Committee on the Peaceful Uses of Outer Space,[2]

Reaffirming the common interest of mankind in furthering the exploration and use of outer space for peaceful purposes,

Recognizing the importance of international co-operation in developing the rule of law in the exploration and peaceful uses of outer space,

Convinced of the need for increased efforts to promote applications of space technology for the benefit of all countries, particularly the developing countries,

Believing that the benefits of space exploration can be extended to States

[1] See *Official Records of the General Assembly, Twenty-fifth Session, Supplement No. 20* (A/8020), annex IV.
[2] *Ibid., Supplement No. 20* (A/8020).

at all stages of economic and scientific development if Member States conduct their space programmes in a manner designed to promote the maximum international co-operation, including the widest possible exchange and practical application of information in this field,

1. *Endorses* the recommendations and decisions contained in the report of the Committee on the Peaceful Uses of Outer Space;

2. *Requests* the Committee on the Peaceful Uses of Outer Space to continue to study questions relative to the definition of outer space and the utilization of outer space and celestial bodies, including various implications of space communications, as well as those comments which may be brought to the attention of the Committee by specialized agencies and the International Atomic Energy Agency as a result of their examination of problems that have arisen or that may arise from the use of outer space in the fields within their competence;

3. *Invites* those States which have not yet become parties to the Treaty on Principles Governing the Activities of States in the Exploration and Use of Outer Space, including the Moon and Other Celestial Bodies and the Agreement on the Rescue of Astronauts, the Return of Astronauts and the Return of Objects Launched into Outer Space to give consideration to ratifying or acceding to those agreements so that they may have the broadest possible effect;

4. *Reaffirms its belief*, as expressed in its resolution 1721 D (XVI) of 20 December 1961, that communication by means of satellites should be available to the nations of the world as soon as practicable on a global and non-discriminatory basis, and recommends that States parties to negotiations regarding international arrangements in the field of satellite communication should constantly bear this principle in mind so that its ultimate realization may be achieved;

5. *Welcomes* the intensified efforts of the Committee on the Peaceful Uses of Outer Space to encourage international programmes to promote such practical applications of space technology as earth resources surveying, for the benefit of both developed and developing countries, and commends to the attention of Member States, specialized agencies and interested United Nations bodies the new programmes and proposals to promote international benefits from space applications noted by the Committee in its report, such as the organization of technical panels, the utilization of internationally sponsored education and training opportunities in the practical applications of space technology and the conduct of experiments in the transfer of space-generated technology to non-space applications;

6. *Takes note* of the recommendation of the Scientific and Technical Sub-Committee of the Committee on the Peaceful Uses of Outer Space that the travel and subsistence of participants in the technical panels mentioned in paragraph 5 above should be funded by their own Governments, but that the United Nations may give timely assistance in exceptional cases within the existing programmes of the United Nations where this appears necessary both to defray costs and to stimulate interest in special areas;

7. *Welcomes* the efforts of Member States to share with other interested Member States the practical benefits which may be derived from their programmes in space technology, including earth resources surveying;

8. *Requests* the Scientific and Technical Sub-Committee, as authorized by the Committee on the Peaceful Uses of Outer Space, to determine at its next session whether, at what time and in what specific frame of reference to convene a working group on earth resources surveying, with special reference to satellites, and in so doing to take into account the importance of appropriate co-ordination with the Committee on Natural Resources, established under Economic and Social Council resolution 1535 (XLIX) of 27 July 1970;

9. *Welcomes* the efforts of Member States to keep the Committee on the Peaceful Uses of Outer Space fully informed of their activities and invites all Member States to do so;

10. *Notes with appreciation* the report of the Expert on Applications of Space Technology concerning the promotion of space applications;[1]

11. *Recalls* the recommendation[2] that Member States give consideration to designating specific offices or individuals, without their Governments, as a point of contact for communications regarding the promotion of the application of space technology and thereafter inform the Secretary-General of such designations, and urges those Member States which have not yet designated a point of contact to do so;

12. *Takes note* of the report provided by the Secretary-General to the Committee on the Peaceful Uses of Outer Space concerning improved co-ordination of Secretariat activities in the field of outer space;[3]

13. *Endorses* the suggestion of the Scientific and Technical Sub-Committee that the Secretary-General should bring to the attention of Member States all relevant documents relating to applications of space technology submitted to the Sub-Committee by Member States, the United Nations, the specialized agencies and other bodies;

14. *Approves* the continuing sponsorship by the United Nations of the Thumba Equatorial Rocket Launching Station and the CELPA Mar del Plata Station and recommends that Member States should give consideration to the use of these facilities for appropriate space research activities;

15. *Notes* that, in accordance with General Assembly resolution 1721 B (XVI) of 20 December 1961, the Secretary-General continues to maintain a public registry of objects launched into orbit or beyond on the basis of information furnished by Member States;

16. *Endorses* the recommendation of the Committee on the Peaceful Uses of Outer Space that the Secretary-General be requested to issue an index of existing international instruments—conventions, treaties and agreements—relating to or bearing upon broadcasting satellite services;

17. *Requests* the specialized agencies and the International Atomic Energy Agency to furnish the Committee on the Peaceful Uses of Outer Space with progress reports on their work in the field of the peaceful uses of outer space, and to examine and report to the Committee on the particular problems which arise or may arise from the use of outer space in the fields within their competence and which should in their opinion be brought to the attention of the Committee;

---

[1] *Ibid.*, annex II.
[2] *Ibid., Twenty-fourth Session, Supplement No. 21* (A/7621), annex II, para. 25.
[3] *Ibid., Twenty-fifth Session, Supplement No. 20* (A/8020), annex III.

18. *Requests* the Committee on the Peaceful Uses of Outer Space to continue its work as set out in the present resolution and in previous resolutions of the General Assembly, and to report to the Assembly at its twenty-sixth session.

1932nd plenary meeting,
16 December 1970.

D

The General Assembly,

Concerned over the devastating and harmful effects of typhoons and storms in various parts of the world, particularly in Asia,

Believing that man's present scientific and technical capabilities that have conquered space could help conquer this environmental scourge,

Recalling its resolutions 1721 (XVI) of 20 December 1961 and 1802 (XVII) of 14 December 1962, and noting the work being undertaken and progress achieved in response to them, as indicated by the World Meteorological Organization in its annual reports to the Committee on the Peaceful Uses of Outer Space,

Noting further the co-ordinating role in this field of the joint Typhoon Committee of the World Meteorological Organization and the Economic Commission for Asia and the Far East, the discussions on this subject held in that forum and the recent decision to transfer the Typhoon Committee secretariat to Manila,

1. *Recommends* that the World Meteorological Organization take, if necessary, further appropriate action for mobilizing capable scientists, technologists and other pertinent resources from any or all nations with a view to obtaining basic meteorological data and discovering ways and means of mitigating the harmful effects of these storms and removing or minimizing their destructive potentials;

2. *Calls upon* Member States to exert efforts within their means to implement fully the World Weather Watch plan of the World Meteorological Organization;

3. *Requests* the World Meteorological Organization to submit a report through the Secretary-General to the Committee on the Peaceful Uses of Outer Space at its next session, and to such other United Nations bodies as may be appropriate, on the steps taken pursuant to the present and other resolutions.

1932nd plenary meeting,
16 December 1970.

2a INTELSAT *Agreement*

ARTICLE I

(k) 'Public telecommunications services' means fixed or mobile telecommunications services which can be provided by satellite and which are available for use by the public, such as telephony, telegraphy, telex, facsimile, data transmission, transmission of radio and television programs between

approved earth stations having access to the INTELSAT space segment for further transmission to the public, and leased circuits for any of these purposes; but excluding those mobile services of a type not provided under the Interim Agreement and the Special Agreement prior to the opening for signature of this Agreement, which are provided through mobile stations operating directly to a satellite which is designed, in whole or in part, to provide services relating to the safety or flight control of aircraft or to aviation or maritime radio navigation;
(l) 'Specialized telecommunications services' means telecommunications services which can be provided by satellite, other than those defined in paragraph (k) of this Article, including, but not limited to, radio navigation services, broadcasting satellite services for reception by the general public, space research services, meteorological services, and earth resource services;
(m) 'Property' includes every subject of whatever nature to which a right of ownership can attach, as well as contractual rights; and
(n) 'Design' and 'development' include research directly related to the purposes of INTELSAT.

ARTICLE II
(*Establishment of* INTELSAT)
(a) With full regard for the principles set forth in the Preamble to this Agreement, the Parties hereby establish the international telecommunications satellite organization 'INTELSAT', the main purpose of which is to continue and carry forward on a definitive basis the design, development, construction, establishment, operation and maintenance of the space segment of the global commercial telecommunications satellite system as established under the provisions of the Interim Agreement and the Special Agreement.
(b) Each State Party shall sign, or shall designate a telecommunications entity, public or private, to sign, the Operating Agreement which shall be concluded in conformity with the provisions of this Agreement and which shall be opened for signature at the same time as this Agreement. Relations between any telecommunications entity, acting as Signatory, and the Party which has designated it shall be governed by applicable domestic law.
(c) Telecommunications administrations and entities may, subject to applicable law, negotiate and enter directly into appropriate traffic agreements with respect to their use of channels of telecommunications provided pursuant to this Agreement and the Operating Agreement, as well as services to be furnished to the public, facilities, divisions of revenue and related business arrangements.

ARTICLE III
(*Scope of* INTELSAT *Activities*)
(a) In continuing and carrying forward on a definitive basis activities concerning the space segment of the global commercial telecommunications satellite system referred to in paragraph (a) of Article II of this Agreement, INTELSAT shall have as its prime objective the provision, on a commercial basis, of the space segment required for international public telecommunica-

tions services of high quality and reliability to be available on a non-discriminatory basis to all areas of the world.

(b) The following shall be considered on the same basis as international public telecommunications services:
  (i) domestic public telecommunications services between areas separated by areas not under the jurisdiction of the State concerned, or between areas separated by the high seas; and
  (ii) domestic public telecommunications services between areas which are not linked by any terrestrial wide-band facilities and which are separated by natural barriers of such an exceptional nature that they impede the viable establishment of terrestrial wide-band facilities between such areas, provided that the Meeting of Signatories, having regard to advice tendered by the Board of Governors, has given the appropriate approval in advance.

(c) The INTELSAT space segment established to meet the prime objective shall also be made available for other domestic public telecommunications services on a non-discriminatory basis to the extent that the ability of INTELSAT to achieve its prime objective is not impaired.

(d) The INTELSAT space segment may also, on request and under appropriate terms and conditions, be utilized for the purpose of specialized telecommunications services, either international or domestic, other than for military purposes, provided that:
  (i) the provision of public telecommunications services is not unfavorably affected thereby; and
  (ii) the arrangements are otherwise acceptable from a technical and economic point of view.

(e) INTELSAT may, on request and under appropriate terms and conditions, provide satellites or associated facilities separate from the INTELSAT space segment for:
  (i) domestic public telecommunications services in territories under the jurisdiction of one or more Parties;
  (ii) international public telecommunications services between or among territories under the jurisdiction of two or more Parties;
  (iii) specialized telecommunications services, other than for military purposes;

provided that the efficient and economic operation of the INTELSAT space segment is not unfavorably affected in any way.

(f) The utilization of the INTELSAT space segment for specialized telecommunications services pursuant to paragraph (d) of this Article, and the provision of satellites or associated facilities separate from the INTELSAT space segment pursuant to paragraph (e) of this Article, shall be covered by contracts entered into between INTELSAT and the applicants concerned. The utilization of INTELSAT space segment facilities for specialized telecommunications services pursuant to paragraph (d) of this Article, and the provision of satellites or associated facilities separate from the INTELSAT space segment for specialized telecommunications services pursuant to subparagraph (e) (iii) of this Article, shall be in accordance with appropriate authorizations, at the planning stage, of the Assembly of Parties pursuant to subparagraph (c) (iv) of Article VII of this Agreement. Where the utilization of INTELSAT space segment facilities for specialized telecommunications services would involve additional costs which result from required modifications to existing or planned INTELSAT space segment facilities, or where the provision of satellites or associated facilities separate from the INTELSAT space segment is sought for specialized telecommunications services as provided for in subparagraph (e) (iii) of this Article, authorization pursuant to subparagraph (c) (iv) of Article VII of this Agreement shall be sought from the Assembly of Parties as soon as the Board of Governors is in a position to advise the Assembly of Parties in detail regarding the estimated cost of the proposal, the benefits to be derived, the technical or other problems involved and the probable effects on present or foreseeable INTELSAT services. Such authorization shall be obtained before the procurement process for the facility or facilities involved is initiated. Before making such authorizations, the Assembly of Parties, in appropriate cases, shall consult or ensure that there has been consultation by INTELSAT with Specialized Agencies of the United Nations directly concerned with the provision of the specialized telecommunications services in question.

ARTICLE IV

(*Juridical Personality*)

(a) INTELSAT shall possess juridical personality. It shall enjoy the full capacity necessary for the exercise of its functions and the achievement of its purposes, including the capacity to:
   (i) conclude agreements with States of international organizations;
   (ii) contract;
   (iii) acquire and dispose of property; and
   (iv) be a party to legal proceedings.
(b) Each Party shall take such action as is necessary within its jurisdiction for the purpose of making effective in terms of its own law the provisions of this Article.

ARTICLE V

(*Financial Principles*)

(a) INTELSAT shall be the owner of the INTELSAT space segment and of all other property acquired by INTELSAT. The financial interest in INTELSAT of each Signatory shall be equal to the amount arrived at by the application of its investment share to the valuation effected pursuant to Article 7 of the Operating Agreement.
(b) Each signatory shall have an investment share corresponding to its percentage of all utilization of the INTELSAT space segment by all Signatories as determined in accordance with the provisions of the Operating Agreement. However, no Signatory, even if its utilization of the INTELSAT space segment is nil, shall have an investment share less than the minimum established in the Operating Agreement.
(c) Each Signatory shall contribute to the capital requirements of INTELSAT, and shall receive capital repayment and compensation for use of capital in accordance with the provisions of the Operating Agreement.
(d) All users of the INTELSAT space segment shall pay utilization charges

determined in accordance with the provisions of this Agreement and the Operating Agreement. The rates of space segment utilization charge for each type of utilization shall be the same for all applicants for space segment capacity for that type of utilization.

(e) The separate satellites and associated facilities referred to in paragraph (e) of Article III of this Agreement may be financed and owned by INTELSAT as part of the INTELSAT space segment upon the unanimous approval of all the Signatories. If such approval is withheld, they shall be separate from the INTELSAT space segment and shall be financed and owned by those requesting them. In this case the financial terms and conditions set by INTELSAT shall be such as to cover fully the costs directly resulting from the design, development, construction and provision of such separate satellites and associated facilities as well as an adequate part of the general and administrative costs of INTELSAT.

2b INTELSAT *Operating Agreement*

ARTICLE 13
(International Telecommunication Union)

In addition to observing the relevant regulations of the International Telecommunication Union, INTELSAT shall, in the design, development, construction and establishment of the INTELSAT space segment and in the procedures established for regulating the operation of the INTELSAT space segment and of the earth stations, give due consideration to the relevant recommendations and procedures of the International Telegraph and Telephone Consultative Committee, the International Radio Consultative Committee and the International Frequency Registration Board.

ARTICLE 14
(Earth Station Approval)

(a) Any application for approval of an earth station to utilize the INTELSAT space segment shall be submitted to INTELSAT by the Signatory designated by the Party in whose territory the earth station is or will be located or, with respect to earth stations located in a territory not under the jurisdiction of a Party, by a duly authorized telecommunications entity.

(b) Failure by the Meeting of Signatories to establish general rules, pursuant to subparagraph (b) (v) of Article VIII of the Agreement, or the Board of Governors to establish criteria and procedures, pursuant to subparagraph (a) (vi) of Article X of the Agreement, for approval of earth stations shall not preclude the Board of Governors from considering or acting upon any application for approval of an earth station to utilize the INTELSAT space segment.

(c) Each Signatory or telecommunications entity referred to in paragraph (a) of this Article shall, with respect to earth stations for which it has submitted an application, be responsible to INTELSAT for compliance of such stations with the rules and standards specified in the document of approval issued to it by INTELSAT, unless, in the case of a Signatory which has submitted an

Appendix of Documents

application, its designating Party assumes such responsibility with respect to all or some of the earth stations not owned or operated by such Signatory.

ARTICLE 15
(Allotment of Space Segment Capacity)
(a) Any application for allotment of INTELSAT space segment capacity shall be submitted to INTELSAT by a Signatory or, in the case of a territory not under the jurisdiction of a Party, by a duly authorized telecommunications entity.
(b) In accordance with the terms and conditions established by the Board of Governors pursuant to Article X of the Agreement, allotment of INTELSAT space segment capacity shall be made to a Signatory or, in the case of a territory not under the jurisdiction of a Party, to the duly authorized telecommunications entity making the application.
(c) Each Signatory or telecommunications entity to which an allotment has been made pursuant to paragraph (b) of this Article shall be responsible for compliance with all the terms and conditions established by INTELSAT with respect to such allotment, unless, in the case of a Signatory which has submitted an application, its designating Party assumes such responsibility for allotments made with respect to all or some of the earth stations not owned or operated by such Signatory.

ARTICLE 16
(Procurement)
(a) All contracts relating to the procurement of goods and services required by INTELSAT shall be awarded in accordance with Article XIII of the Agreement, Article 17 of this Operating Agreement and the procedures, regulations, terms and conditions established by the Board of Governors pursuant to the provisions of the Agreement and this Operating Agreement. The services to which this Article refers are those provided by juridical persons.
(b) The approval of the Board of Governors shall be required before:
  (i) the issuing of requests for proposals or invitations to tender for contracts which are expected to exceed 500,000 U.S. dollars in value;
  (ii) the awarding of any contract to a value exceeding 500,000 U.S. dollars.
(c) In any of the following circumstances, the Board of Governors may decide to procure goods and services otherwise than on the basis of responses to open international invitations to tender:
  (i) where the estimated value of the contract does not exceed 50,000 U.S. dollars or any such higher amount as the Meeting of Signatories may decide in the light of proposals by the Board of Governors;
  (ii) where procurement is required urgently to meet an emergency situation involving the operational viability of the INTELSAT space segment;
  (iii) where the requirement is of a predominantly administrative nature best suited to local procurement; and
  (iv) where there is only one source of supply to a specification which is necessary to meet the requirements of INTELSAT or where the sources

of supply are so severely restricted in number that it would be neither feasible nor in the best interest of INTELSAT to incur the expenditure and time involved in open international tender, provided that where there is more than one source they will all have the opportunity to bid on an equal basis.

(d) The procedures, regulations, terms and conditions referred to in paragraph (a) of this Article shall provide for the supply of full and timely information to the Board of Governors. Upon request from any Governor, the Board of Governors shall be able to obtain, with respect to all contracts, any information necessary to enable that Governor to discharge his responsibilities as a Governor.

ARTICLE 17
(Inventions and Technical Information)

(a) INTELSAT, in connection with any work performed by it or on its behalf, shall acquire in inventions and technical information those rights, but no more than those rights, necessary in the common interests of INTELSAT and the Signatories in their capacity as such. In the case of work done under contract, any such rights obtained shall be on a non-exclusive basis.

(b) For the purposes of paragraph (a) of this Article, INTELSAT, taking into account its principles and objectives, the rights and obligations of the Parties and Signatories under the Agreement and this Operating Agreement and generally accepted industrial practices, shall, in connection with any work performed by it or on its behalf involving a significant element of study, research or development, ensure for itself:

  (i) the right without payment to have disclosed to it all inventions and technical information generated by work performed by it or on its behalf;

  (ii) the right to disclose and have disclosed to Signatories and others within the jurisdiction of any Party and to use and authorize and have authorized Signatories and such others to use such inventions and technical information:

   (A) without payment, in connection with the INTELSAT space segment and any earth station operating in conjunction therewith, and

   (B) for any other purpose, on fair and reasonable terms and conditions to be settled between Signatories or others within the jurisdiction of any Party and the owner or originator of such inventions and technical information or any other duly authorized entity or person having a property interest therein.

(c) In the case of work done under contract, the implementation of paragraph (b) of this Article shall be based on the retention by contractors of ownership of rights in inventions and technical information generated by them.

(d) INTELSAT shall also ensure for itself the right, on fair and reasonable terms and conditions, to disclose and have disclosed to Signatories and others within the jurisdiction of any Party, and to use and authorize and have authorized Signatories and such others to use, inventions and technical information directly utilized in the execution of work performed on its behalf but not included in paragraph (b) of this Article, to the extent that the person who has performed such work is entitled to grant such right and to the extent that such disclosure and use is necessary for the effective exercise of rights obtained pursuant to paragraph (b) of this Article.

(e) The Board of Governors may, in individual cases, where exceptional circumstances warrant, approve a deviation from the policies described in subparagraph (b) (ii) and paragraph (d) of this Article where in the course of negotiations it is demonstrated to the Board of Governors that failure to deviate would be detrimental to the interests of INTELSAT and, in the case of subparagraph (b) (ii), that adherence to these policies would be incompatible with prior contractual obligations entered into in good faith by a prospective contractor with a third party.

(f) The Board of Governors may also, in individual cases, where exceptional circumstances warrant, approve a deviation from the policy prescribed in paragraph (c) of this Article where all of the following conditions are met:
- (i) it is demonstrated to the Board of Governors that failure to deviate would be detrimental to the interests of INTELSAT,
- (ii) it is determined by the Board of Governors that INTELSAT should be able to ensure patent protection in any country and
- (iii) where, and to the extent that, the contractor is unable or unwilling to ensure such protection on a timely basis.

(g) In determining whether and in what form to approve any deviation pursuant to paragraphs (e) and (f) of this Article, the Board of Governors shall take into account the interests of INTELSAT and all Signatories and the estimated financial benefits to INTELSAT resulting from such deviation.

(h) With respect to inventions and technical information in which rights were acquired under the Interim Agreement and the Special Agreement, or are acquired under the Agreement and this Operating Agreement other than pursuant to paragraph (b) of this Article, INTELSAT, to the extent that it has the right to do so, shall upon request:
- (i) disclose or have disclosed such inventions and technical information to any Signatory, subject to reimbursement of any payment made by or required of INTELSAT in respect of the exercise of such right of disclosure;
- (ii) make available to any Signatory the right to disclose or have disclosed to others within the jurisdiction of any Party and to use and authorize or have authorized such others to use such inventions and technical information:
  - (A) without payment, in connection with the INTELSAT space segment or any earth station operating in conjunction therewith, and
  - (B) for any other purpose, on fair and reasonable terms and conditions to be settled between Signatories or others within the jurisdiction of any Party and INTELSAT or the owner or originator of such inventions and technical information or any other duly authorized entity or person having a property interest therein, and subject to reimbursement of any payment made by or required of INTELSAT in respect of the exercise of such rights.

(i) To the extent that INTELSAT acquires the right pursuant to sub-paragraph (b) (i) of this Article to have inventions and technical information disclosed to it, it shall keep each Signatory which so requests informed of the availability and general nature of such inventions and technical information. To the extent that INTELSAT acquires rights pursuant to the provisions of this Article to make inventions and technical information available to Signatories and others in the jurisdiction of Parties, it shall make such rights available upon request to any Signatory or its designee.

(j) The disclosure and use, and the terms and conditions of disclosure and use, of all inventions and technical information in which INTELSAT has acquired any rights shall be on a non-discriminatory basis with respect to all Signatories and their designees.

ARTICLE 18

(*Liability*)

(a) Neither INTELSAT nor any Signatory, in its capacity as such, nor any director, officer or employee of any of them nor any representative to any organ of INTELSAT acting in the performance of their functions and within the scope of their authority, shall be liable to, nor shall any claim be made against any of them by, any Signatory or INTELSAT for loss or damage sustained by reason of any unavailability, delay or faultiness of telecommunications services provided or to be provided pursuant to the Agreement or this Operating Agreement.

(b) If INTELSAT or any Signatory, in its capacity as such, is required, by reason of a binding decision rendered by a competent tribunal or as a result of a settlement agreed to or concurred in by the Board of Governors, to pay any claim, including any costs and expenses associated therewith, which arises out of any activity conducted or authorized by INTELSAT pursuant to the Agreement or to this Operating Agreement, to the extent that the claim is not satisfied through indemnification, insurance or other financial arrangements, the Signatories shall, notwithstanding any ceiling established by or pursuant to Article 5 of this Operating Agreement, pay to INTELSAT the amount unsatisfied on such claim in proportion to their respective investment shares as of the date the payment by INTELSAT of such claim is due.

(c) If such a claim is asserted against a Signatory, that Signatory, as a condition of payment by INTELSAT of the claim pursuant to paragraph (b) of this Article, shall without delay provide INTELSAT with notice thereof, and shall afford INTELSAT the opportunity to advise and recommend on or to conduct the defense or other disposition of the claim and, to the extent permitted by the laws of the jurisdiction in which the claim is brought, to become a party to the proceeding either with such Signatory or in substitution for it.

3
Declaration by the International Conference of Broadcasting Unions on Communication Satellites
Rome, 6–11 March 1972,

attended by ABU (Asia and the Pacific), AIR (the Americas), ASBU (Arab States), EBU (Western Europe and Mediterranean countries), OTI (Ibero-America), URTNA (Africa), and representatives of North American organizations (ABC, CBC, CBS, CPB, NAEB, NBC and USIA)

Considering that communication satellites have made available to the broadcasters of the world an unprecedented medium for the dissemination of information, education and culture,

Being aware, in the use of this new medium, of the rules laid down by the United Nations and the Allocations Rules of the International Telecommunication Union, and noting that UNESCO proposes to adopt a Declaration of Guiding Principles at its 17th General Conference,

The associations and organizations enumerated above have agreed subject to ratification by their respective governing bodies, on the following declaration:

1. The associations and organizations enumerated above (hereinafter called the 'broadcasters') recognize the increased responsibilities which they bear through the opportunity offered to them for the use of communication satellites to have access to programme sources throughout the world and to distribute their programmes to worldwide audiences.

2. In fulfilment of these responsibilities broadcasters shall, by the widest possible dissemination of information, education and culture throughout the world, utilize the new medium of distribution to promote peace and mutual understanding.

3. In the field of information, the broadcasters declare their intention to co-operate and to provide all possible mutual assistance for the supply of news material on events throughout the world.

4. In the field of education, the broadcasters declare their intention to co-operate and to provide all possible mutual assistance, so that they may, if they deem necessary, broadcast in their programmes all possible material designed to improve or complement school or post-school education.

5. In the cultural field, the broadcasters declare their intention to co-operate and to provide all possible mutual assistance, so that they may, in accordance with their conceptions, broadcast in their programmes works drawn from the most varied cultural heritages.

6. For the attainment of the above-mentioned objectives, the broadcasters shall, to the extent it lies in their power, provide mutual assistance in facilitating each other's access to the Earth stations situated in their respective territories, regardless of who may be the recipient(s) of the programme thus transmitted.

7. Subject to the provisions of paragraph 11 below, the broadcasters mutually recognize, as far as they are concerned, their freedom to use satellites for the transmission and reception of programmes, reception

however remaining subject to the prior consent of the originating broadcaster, which may need to comply with legal restrictions.

8. The broadcasters declare their intention to promote professional co-operation among themselves with a view to optimum attainment of the objectives set forth herein.

9. Such co-operation shall as a matter of priority embrace the broadcasters of developing countries and their associations.

10. The co-operation shall include inter alia joint work towards
   (i) the setting up of regional technical and programme co-ordination centres,
   (ii) adopting the procedures to be followed in assuring an exchange of information on topics of common interest, particularly programmes,
   (iii) devising methods for the training of personnel in the various fields specific to satellite transmissions,
   (iv) devising subsequent professional arrangements in other areas, particularly the technical and legal fields.

11. The broadcasters note that the reception within one state of a programme distributed by direct broadcast satellite from another state may be regulated by an agreement between the appropriate bodies in the states concerned.[1]

SELECT DOCUMENTS OF WORLD ADMINISTRATIVE
RADIO CONFERENCE 1971

Revision of Article 7 of the Radio Regulations

Spa2   **Section IA. Broadcasting-Satellite Service**

428A   §2A. In devising the characteristics of a space station in the broad-
Spa2   casting-satellite service, all technical means available shall be used to reduce, to the maximum extent practicable, the radiation over the territory of other countries unless an agreement has been previously reached with such countries.

Spa2   **Section IX. Space Radiocommunication Services**

Cessation of Emissions

470V   §24. Space stations shall be fitted with devices to ensure immediate
Spa2   cessation of their radio emissions by telecommand, whenever such cessation is required under the provisions of these Regulations.

---

[1] The North American organizations ABC, CBS and NBC felt this article was incomplete unless the following phrase was added after the word 'concerned': '. . . if such agreement is consistent with the fundamental law of each state concerned'.

RESOLUTION No. Spa2—1

Relating to the Use by all Countries, with equal Rights, of Frequency Bands for Space Radiocommunication Services

The World Administrative Radio Conference for Space Telecommunications (Geneva, 1971),

considering

that all countries have equal rights in the use of both the radio frequencies allocated to various space radiocommunication services and the geostationary satellite orbit for these services;

taking into account

that the radio frequency spectrum and the geostationary satellite orbit are limited natural resources and should be most effectively and economically used;

having in mind

that the use of the allocated frequency bands and fixed positions in the geostationary satellite orbit by individual countries or groups of countries can start at various dates depending on requirements and readiness of technical facilities of countries;

resolves

1. that the registration with the ITU. of frequency assignments for space radiocommunication services and their use should not provide any permanent priority for any individual country or groups of countries and should not create an obstacle to the establishment of space systems by other countries;

2. that, accordingly, a country or a group of countries having registered with the ITU frequencies for their space radiocommunication services should take all practicable measures to realize the possibility of the use of new space systems by other countries or groups of countries so desiring;

3. that the provisions contained in paragraphs 1 and 2 of this Resolution should be taken into account by the administrations and the permanent organs of the Union.

RESOLUTION No. Spa2—2

Relating to the Establishment of Agreements and Associated Plans for the Broadcasting-Satellite Service

The World Administrative Radio Conference for Space Telecommunications (Geneva, 1971),

considering

a) that it is important to make the best possible use of the geostationary-satellite orbit and of the frequency bands allocated to the broadcasting-satellite service;

b) that the great number of receiving installations using such directional antennae as could be set up for a broadcasting-satellite service may be an obstacle to changing the location of space stations in that service on the geostationary-satellite orbit, from the date of their bringing into use;

c) that satellite broadcasts may create harmful interference over a large area of the Earth's surface;

d) that the other services with allocations in the same band need to use the band before the broadcasting-satellite service is set up;

resolves

1. that stations in the broadcasting-satellite service shall be established and operated in accordance with agreements and associated plans adopted by World or Regional Administrative Conferences, as the case may be, in which all the administrations concerned and the administrations whose services are liable to be affected may participate;

2. that the Administrative Council be requested to examine as soon as possible the question of a World Administrative Conference, and/or Regional Administrative Conferences as required, with a view to fixing suitable dates, places and agenda;

3. that during the period before the entry into force of such agreements and associated plans the administrations and the IFRB shall apply the procedure contained in Resolution No. **Spa2**—3.

RECOMMENDATION No. **Spa2**—1

Relating to the Examination by World Administrative Radio Conferences of the Situation with Regard to Occupation of the Frequency Spectrum in Space Radiocommunications

The World Administrative Radio Conference for Space Telecommunications (Geneva, 1971),

considering

a) that the frequency bands available for space applications are limited in number and size;

b) that the possible positions for a satellite whose main purpose is to establish telecommunication links are limited in number and that certain positions are more favourable than others for certain links;

c) that all administrations should be enabled to establish space links which they deem necessary;

d) that the scale and cost of space networks or systems are such that their operation and development must be hindered as little as possible;

e) that technology is steadily and rapidly evolving and that the best possible use should be made of resources in space radiocommunications;

f) that administrations should ensure that frequency assignments for space applications are utilized in the most efficient manner possible consistent with developing technology and that such assignments are relinquished when no longer in use;

g)   that despite the provisions of Article 9A of the Radio Regulations and the principles adopted by this Conference, which provide for full consultation and co-ordination between administrations with a view to the optimum accommodation of all space systems, it is possible that as the use of frequencies and orbital positions increases, administrations may encounter undue difficulty in one or more frequency bands in meeting their requirements for space radiocommunication;

recommends

that the next appropriate World Administrative Radio Conference be empowered to deal with the situation described in Considering *g)*, if it arises;

invites

the Administrative Council, in the event of such a situation arising, to include in the agenda for the next appropriate World Administrative Radio Conference specific provisions enabling it to examine all aspects of the use of the frequency band(s) concerned including, *inter alia*, the relevant frequency assignments recorded in the Master International Frequency Register and to find a solution to the problem.

List of Terms

Bandwidth	Largeur de bande	Amplitud de banda
Beam	Faisceau	Haz
Cable	Fil	Linea alambrica
Channel	Voie	Canale
Communications	Télécommunications	Comunicaciónes
Community broadcast	Diffusion communauté	Transmisión a una comunidad
Data transmission	Transmission de données	Transmisión de los datos
Direct broadcast	Diffusion directe	Emisión directa
Earth station	Installation sol	Estación terrestre
Launcher/Rocket	Lanceur/Fusée porteuse	Vehiculo de lanzamiento/Cohete portador
Microwave	Hyperfréquence	Microonda
Monitoring	Contrôle	Control
National TV system	Réseau national de TV	Sistema de TV de amplitud nacional
Outer space	Espace extra-atmosphérique	Espacio ultraterrestre
Point-to-point	Point à point	Punto a punto
Radio interference	Brouillage	
Receiver	Récepteur	Receptor
Relay/Rebroadcast Station	Station de retransmission	Estación de retransmisión
Satellite broadcast	Diffusion par satellite	Transmisión mediante satelite
Signal	Signal	Señal
Synchronous orbit	Orbite d'attente	Orbita sincrónica
Technical 'spin-off'	Retombée technique	Repercusión en la técnica

Select Bibliography

The literature is becoming so abundant that only a limited number of publications of particular interest or relevance in this context have been listed.

1. BIBLIOGRAPHIES
2. OFFICIAL DOCUMENTS OF INTERNATIONAL ORGANIZATIONS
3. GENERAL WORKS
4. SATELLITE COMMUNICATIONS and SATELLITE BROADCASTING

1. *Bibliographies*

EDSAT CENTER: *The Educational and Social Use of Communications Satellites.* Edsat Program, University of Wisconsin, Madison, Wisconsin, 1970.
────*Legal and Political Aspects of Satellite Telecommunications.* Edsat Center, University of Wisconsin, Madison, Wisconsin, 1971.
INTERNATIONAL INSTITUTE OF SPACE LAW. *Worldwide Bibliography for the years.* International Astronautic Federation, Paris.
UNESCO. An annotated bibliography of Unesco publications and documents dealing with space communications 1953–70. Unesco, doc. no. COM/WS/166, Paris, 1971.
UNITED NATIONS. *International Space Bibliography.* New York, 1966.

2. *Official Documents of International Organizations*

ITU
Extraordinary Administrative Radio Conference to Allocate Frequency Bands for Space Radio Communication Purposes, Geneva 1963. *Final Acts.* Geneva, 1963.
International Telecommunications Convention, Montreux 1965. Geneva, 1965.
Radio Regulations, Geneva 1959. New edition, Geneva, 1968,
Telecommunications and Peaceful Uses of Outer Space, *Reports*: 1st, 1962; 2nd, 1963; 3rd 1964; 4th, 1965; 5th 1966; 6th, 1967; 7th, 1968; 8th, 1969; 9th, 1970; 10th, 1971; 11th, 1972.
World Administrative Radio Conference for Space Telecommunications, Geneva 1971. *Final Acts*, Geneva, 1971.

UNITED NATIONS
GA resolutions relating to problems of communications satellites and direct broadcast satellites:
111 (II) 3 Nov. 1947: Condemnation of war propaganda.
424 (V) 14 Dec. 1950: Freedom of Information: Interference with Radio signals.
841 (X) 17 Dec. 1954: Internat. Convention concerning the use of Broadcasting in the Cause of Peace. Geneva, 1936.

1772 (XIV)   31 Dec. 1959: International Co-operation in the Peaceful Uses of Outer Space.
1721 (XVI)   20 Dec. 1961: International Co-operation in the Peaceful Uses of Outer Space.
1802 (XVII)   14 Dec. 1962: International Co-operation in the Peaceful Uses of Outer Space.
1962 (XVIII)   13 Dec. 1963: Declaration of Legal Principles Governing the Activities of States in the Exploration and Use of Outer Space.
1963 (XIV)   13 Dec. 1963: International Co-operation in the Peaceful Uses of Outer Space.
2222 (XXI)   19 Dec. 1966: Treaty on Principles Governing the Activity of States in the Exploration and Use of Outer Space, including the Moon and other celestial bodies.
2448 (XXIII)   19 Dec. 1968: Freedom of Information.
2453 (XXIII)   20 Dec. 1968: International Co-operation in the Peaceful Uses of Outer Space.
2601 (XXIV)   16 Dec. 1969: International Co-operation in the Peaceful Uses of Outer Space.
2453 (XXIII)   Dec. 1969: International Co-operation in the Peaceful Uses of Outer Space.
2733 (XXV)   Dec. 1970: International Co-operation in the Peaceful Uses of Outer Space.
Committee on the Peaceful Uses of Outer Space, Working Group on Direct Broadcast Satellites
1st session, New York, 11–20 Feb. 1969. Report (A/AC.105/51); working papers from Australia, Canada-Sweden (A/AC.105/50), ITU (A/AC.105/52) and Unesco.
2nd session, Geneva, 29 July–7 Aug. 1969. Report (A/AC.105/66); working papers from Argentina (A/AC.105/WG. 3/WPI), Australia (A/AC.105/63), Canada-Sweden (A/AC.105/59), Czechoslovakia (A/AC.105/61), France (A/AC.105/62), Mexico (A/AC.105/64), United Kingdom (A/AC.105/65), and Unesco (A/AC.105/60)
3rd session, New York, 11 May 1970. Report (A/AC.105/83); working papers from Canada-Sweden (A/AC.105/WG3/L1), France (A/AC.105/WG3/CRP2), and USSR (A/AC.105/WG3/CRP1). Comments received from governments, specialized agencies, and other competent international bodies (A/AC.105/79)
Space Exploration and Applications; Papers presented at the United Nations Conference on the Exploration and Peaceful Uses of Outer Space, Vienna 14–27 August, 1968, vols I & II, New York, 1969.

UNESCO
General Conference:
resolutions adopted at 11th session, 1960; 12th session, 1962; 13th session, 1964; 14th session, 1966; 15th session, 1968; 16th session, 1970.
Broadcasting from space. Reports and papers on Mass Communication, No. 60. Paris, Unesco.
GJESDAL, Tor. 'Unesco's programme in space communication'. In *Unesco Chronicle* (Paris), XVI/11, Nov. 1970.
Aide-Memoire on frequency allocations for space communication; the

Select Bibliography

significance for Unesco's aims of the WARC-ST. Paris, Unesco, 1970.

Long-term programme for the use of space communication: proposals by the Director-General. Paris, Unesco, 1966 (Unesco 14 C/25).

Meeting of Experts on the Use of Space Communication by the Mass Media, Paris, 1965, Unesco 1966 (UNESCO/MC/52).

Communication in the space age; the use of satellites by the mass media. Paris, Unesco, 1968.

Meeting of Experts on the Use of Space Communication for Broadcasting, Paris, 1968. Report. Paris, Unesco, 1968 (COM/CS/68/1/7).

Meeting of Panel of Officers for the Meeting of Experts on the Use of Space Communications by the Mass Media, Paris, 1966. Report. Paris, Unesco, 1966 (UNESCO/SPACE/COM/PANEL 1/4).

Meeting of Unesco Advisory Panel on Space Communication, Paris, 2nd session, 1967. Report. Paris, Unesco, 1967 (COM/CS/31/1).

Meeting of Unesco Advisory Panel on Space Communication, Paris, 3rd session, 1967. Report. Paris, Unesco, 1968 (COM/CS/164/5).

Meeting of Unesco Advisory Panel on Space Communication, Stockholm 4th session, 1968. Report. Paris, Unesco 1968 (COM/SPACE/PANEL/IV/3).

Meeting of Unesco Panel of Consultants on Space Communication, Paris. 5th session, 1969. Report. Paris, Unesco (COM/CONF/1/2).

Meeting of Unesco Panel of Consultants on Space Communication, Paris, 6th session, 1970. Report, Paris, Unesco 1970 (COM/SPACE/PANEL/VI/3).

Ploman, Edward. A guide to satellite communication. Paris, 1972 (Reports and Papers on Mass Communication No. 66).

Schramm, Wilbur. Communication satellites for education, science and culture. Paris, 1968 (Reports and Papers on Mass Communication No. 53).

Unesco Expert Mission to Alaska to study the implications of satellite communication for education, August–September 1970. (2198/BMS,RD/MC).

Unesco Expert Mission to Brazil to make a preparatory study of the use of satellite communication for education and national development, May–June 1968. Report. Paris, Unesco, 1968. (COM/WS/86).

Unesco Expert Mission to India to make a preparatory study of a pilot project in the use of satellite communication for national development purposes, November–December 1967. Report. Paris, Unesco, 1968. (COM/WS/51).

Unesco Expert Mission to Pakistan to study the possibilities of using space communication for education and development, February–March 1969. Report. Paris, Unesco, 1969. (COM/WS/142).

Unesco Expert Mission to South America to make a preparatory study of the use of satellite communication for education and national development, August–September 1969. Report. Paris, Unesco, 1970. (1903/BMS.RD/COM).

3. General Works

AFSHAR, H. K. *The innovative consequences of space technology and the problems of the developing countries.* 1971.

BERRADA, A. *Frequencies for broadcasting satellites.* London, IBI, 1972.

BLAMONT, J. E. Les moyens spatiaux au service de l'éducation et de la formation des masses. *Recherche spatiale*, no. 10, Oct. 1968.

BÖELLE, G. and others. *L'utilisation de satellites de diffusion directe*. Paris, Presses Universitaires de France, 1970.

BROWN, G. M. *Space radio communication*. Amsterdam, Elsevier, 1962.

Centre National de la Recherche Scientifique, Groupe de travail sur le droit de l'espace (ouvrage collectif). *Les télécommunications par satellites, aspects juridiques* . . . Paris, Cujas, 1968.

CHAUMERON, J. L'évolution des satellites de télécommunications. *R. française d'astronautique*, no. 6, 1964.

CHAYES, A. & L. CHAZEN. Policy problems in direct broadcasting from satellites. *Stanford J. Internat. Study*, June 1970.

CLARKE, ARTHUR C. *Voices from the sky*. London, Gollancz, 1966.

COCCA, A. A. *Regimen legal de las communicaciones espaciales*. Cordoba, 1966.

——Promises and menaces emerging from direct broadcast by satellites. Paper presented at Conference on World Peace through the Rule of Law, Bangkok, Sept. 1969.

CODDING, G. A. J. *La radiodiffusion dans le monde*. Paris, Unesco, 1959.

DALFEN, C. M. Direct satellite broadcasting: towards international agreements to transcend and marshal the political realities. *Toronto Law J.*, 1970.

——The Telesat Canada domestic communications satellite system. *Stanford J. Internat. Stud.*, June 1970.

DEBBASCH, C. *Traité du droit de la radiodiffusion, radio et télévision*. Paris, Lib. Gen. de Droit et de Jurisprudence, 1967.

——*Le droit de la radio et de la télévision*. Paris, Presses Universitaires de France, 1969.

DELORME, J. La télévision par satellites: force de persuasion. *R. politique et parlementaire*. Mar., 1966

DOYLE, S. E. Communication satellites, international organization for development and control. *California Law R.*, May 1967.

EMERY, W. B. *National and International Systems of Broadcasting*. Michigan, State Univ. Press, East Lansing, 1969.

ESTEP, S. D. International Lawmakers in a Technological World, Space communications and nuclear energy. *George Washington Law R.*, Oct. 1964.

EVENSEN, J. *Aspects of international law relating to modern radio communications*.

FRENKEL, H. M. & A. E. *World peace via satellite communications*. New York, Telecommunications Research Associates, 1965.

GARDNER, R. N. Space communications: a new instrument for world order. *War/Peace Report*, Oct. 1968.

GESTON, M. S. Direct broadcasting from satellite, the case for regulation. *New York Univ. J. Internat. Law and Politics*, Spring 1970.

HAVILAND, R. P. Space broadcasting—how, when and why. *Bull. Atomic Scientists*, Mar. 1969.

KLEIN, B. M. and others. *Communications satellites and public policy; an introductory report*. Santa Monica, RAND Corp., 1961.

LEINWOLL, S. Direct broadcasting from earth satellites, in *Space communications*. New York, Rider, 1964.

LEIVE, D. M. *International communications and international law: the regulation of the radio spectrum*. Leyden, Sijthoff, 1970.

MICKELSON, S. Communications by satellites. *Foreign Affairs*, Oct. 1969.
MCWHINNEY, E. *The international law of communications*. Leyden, Sijthoff, 1971.
NAMUROIS, A. *The Organization of Broadcasting*. Geneva, EBV.
PAULU, B. *Radio and television broadcasting on the European continent*. Minneapolis, Univ. of Minnesota Press, 1967.
ROSTOW REPORT. President's Task Force on Communications Policy, *Final report*. Washington, USGPO, 1968.
RYBAKOV, Y. M. International Legal Co-operation in Space. *Sovetskoye Gosudarstvo Pravo*, No. 2, 1970.
RYDBECK, O. & PLOMAN, E. *Les communications spatiales et la radiodiffusion*. Monographies juridiques et administratives de l'UER, 1969.
—— Les organismes européens de radiodiffusion et de télévision face à la technologie des satellites. *J. des télécommunications*, Feb. 1969.
SHILLINGLAW, T. L. The Soviet Union and international satellite telecommunications. *Stanford J. Internat. Stud.*, June 1970.
SMITH, D. D. *International telecommunication control*. Leyden, Sijthoff, 1969.
STOLUSKY, W. G. Unauthorized interception of Space oriented telecommunications. *Federal Bar J.*, Fall 1965.
STRASCHNOV, G. *Le droit d'auteur et les droits connexes en radiodiffusion*. Brussels, Bruylant, 1950.
—— Legal protection of television broadcasts transmitted via satellite—against their use without the permission of the originating organization. *Bulletin of the copyright society of the USA*, Oct. 1969.
TASSIN, J. *Vers l'Europe spatiale*. Paris, Denoel, 1970.
TERROU, F. *L'information*. 2nd ed. Paris, Presses Universitaires de France, 1965.
THOMAS, G. L. Approaches to controlling propaganda and spillover from Direct-Broadcast Satellites. *Stanford J. Internat. Stud.*, June 1970.
Twentieth Century Fund Task Force on International Satellite Communications. *The Future of Satellite Communications, Resource Management and the Needs of Nations*. New York, Twentieth Century Fund, 1970.
WALTERS, E. N. Perspectives in the emerging Law of Satellite Communications. *Stanford J. Internat. Stud.*, June 1970.

4. *Satellite Communications and Satellite Broadcasting*

a) A number of institutions and organizations publish studies, articles, etc., concerning various aspects of satellite communications relevant in this context, either in regular publications, as conference reports or otherwise; as examples could be mentioned
 European Broadcasting Union: *EBU Review*, Geneva
 International Telecommunication Union: *International Telecommunication Journal*, Geneva
 International Astronautical Federation, Paris
 International Institute of Space Law, Paris
 American Institute of Aeronautics and Astronautics, New York
 Centre National de la Recherche Scientifique, Paris
 Reports of US Congress
 National Aeronautics and Space Administration (NASA), Washington, DC.